Updating to Remain the Same

Updating to Remain the Same

Habitual New Media

Wendy Hui Kyong Chun

The MIT Press
Cambridge, Massachusetts
London, England

First MIT Press new paperback edition, 2017

This book was set in Stone Sans and Stone Serif by Toppan Best-set Premedia Limited. Printed and bound in the United States of America.

Library of Congress Cataloging-in-Publication Data

Names: Chun, Wendy Hui Kyong, 1969- author.
Title: Updating to remain the same : habitual new media / Chun, Wendy Hui Kyong.
Description: Cambridge, MA : The MIT Press, 2016. | Includes bibliographical references and index.
Identifiers: LCCN 2015039932 | ISBN 9780262034494 (hardcover : alk. paper)
Subjects: LCSH: Internet--Social aspects. | Information society. | Mass media and technology. | Digital media--Social aspects.
Classification: LCC HM851 .C486 2016 | DDC 302.23/1--dc23 LC record available at http://lccn.loc.gov/2015039932

ISBN: 978-0-262-03449-4 (alk. paper: hc.)—978-0-262-53472-7 (pb.)

10 9 8 7 6 5 4

For the members of the Center for Digital Culture in Lüneburg, who made this book possible.

For my mother and father, who have made everything possible.

Contents

Preface: The Wonderful Creepiness of New Media

Has the Internet destroyed the world or made it a better place? Does it foster democracy or total surveillance? Community or isolation? Information or pornography? Well-adjusted citizens or homicidal psychopaths?

These questions have been posed and answered over and over again. And they will continue to be so, unless we change our perspective and realize—at the very least—that the Internet is not just one thing. So, rather than endless debate over whether the Internet is good or bad, this book asks: Why does the Internet evoke such contradictory passions? Its answer: new media are so powerful because they mess with the distinction between publicity and privacy, gossip and political speech, surveillance and entertainment, intimacy and work, hype and reality. **New media are wonderfully creepy.** They are endlessly fascinating yet boring, addictive yet revolting, banal yet revolutionary.

To understand new media's wonderful creepiness, we need to disabuse ourselves of several assumptions, most importantly that there exists a "natural" relationship between technology and (the lack of) freedom. Remarkably, the image of the Internet has shifted radically from the mid to late 1990s, when it was seen as "cyberspace," an anonymous and empowering space of freedom in which no one knew if you were a dog, to the mid to late 2010s, when the Internet was commonly conceived of as a space of total surveillance or as a privatized space of social media. In both cases, knowing who was a dog and who was not was key (see figure 0a.2).

So, what happened? Did the technology itself change radically? Yes and no. Internet protocol is still basically the same, even if routing, addressing, and storage protocols have changed; but there is no simple relationship between these technical changes and the different imaginaries of the Internet. There is a gap between our perception of communications technologies and their habitual operations. Wireless networking cards, for example, download all the packets they can and then delete what is not directly

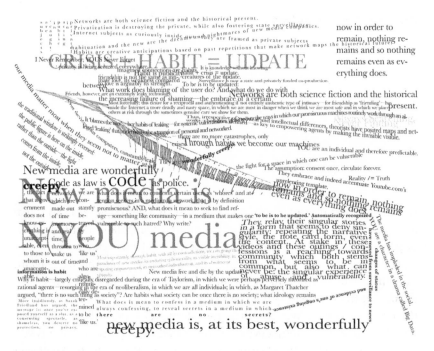

Figure 0a.1
Habitual New Media word cloud. Image by Thomas Pringle.

addressed to them. Users constantly download their neighbor's traffic. Given this, why did we ever imagine the Internet—which is, at its base, a control protocol—to be an anonymous space of freedom? Why are networked devices described as "personal," when they are so chatty and promiscuous? Further, given the ephemerality of digital information, how has electronic memory become conflated with storage? It takes a lot of work—at the very least the U.S. National Security Agency, social media sites that foster cross-platform "unique identifiers," and vast server farms that contribute to global climate change—to make the Internet the basis for worldwide surveillance. If the phrase "once it's there, it's there to stay" makes any sense, it is because surveillance is now co-produced transnationally by states and private corporations.

To understand the power of our imagined technologies and networks, this book focuses on how they ground and foster habits of using. By focusing on habits—things that remain by disappearing from consciousness—it reveals the creepier, slower, more unnerving time of "new media." Interrogating the hype around "disruption," it shows how so-called obsolescent

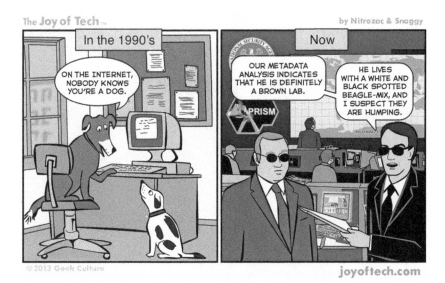

Figure 0a.2
"On the Internet, nobody knows you're a dog." Nitrozac & Snaggy, *Joy of Tech* (2013), http://www.geekculture.com/joyoftech/joyarchives/1862.html. Reprinted with permission.

media remain in users' bodies. If users now "curate" their lives, it is because their bodies have become archives. Habit also gets us thinking about the complex relationship between what is allegedly public and private, intimate and social. Habits are things we learn from others, and they make us 'like' others. At the same time, though, they are deeply personal. Working with this contradiction, this book explores the extent to which habits are what society can be in the era of neoliberalism, an epoch that emphasizes individual empowerment and difference. Neoliberal subjects are constantly encouraged to change their habits—rather than society and institutions—in order to become happier, more productive people; to recycle rather than regulate in order to save the world. However, instead of simply bemoaning this situation, this book also asks: Can we—by inhabiting the habitual—change society? Further, rather than pushing for a privacy that is no privacy—a security that fosters insecurity—what would happen if we demanded more rigorous public rights? If we fought for the right to be exposed—to take risks and to be in public—and not be attacked?

Acknowledgments

Research for this book was supported by several grants, leaves, and fellowships. I would like to thank the Institute for Advanced Study (IAS) at Princeton, the Center for Digital Cultures (CDC) and the Digital Cultures Research Lab (DCRL) both in Lüneburg, and Brown University for their generous support.

I would also like to thank the many people who have read drafts of this book for their invaluable feedback, in particular the members of the writing groups at the CDC and DCRL, and attendees of seminars at the University of Amsterdam, Brock University, Pitzer College, the Copenhagen Business School, and the University of California at Davis. I am also grateful to the many colleagues and friends who have inspired me at Brown, in Lüneburg (Clemens Apprich, Goetz Bachmann, Armin Beverungen, Timon Beyes, Mercedez Bunz, Paul Feigelfeld, Sam Gupta, Claus Pias, Oliver Schultz, Nishant Shah, Madoka Takashiro, Martin and Carmen Warnke, but really everyone), and in Princeton (Danielle Allen, Elizabeth Bernstein, Sherine Hamdy, Amy Kaplan, Karuna Mantena, Camille Robcis, Judith Surkis, and Joan Scott), as well as my constant sources of brilliance: Liz Canner, Florian Cramer, Mary Ann Doane, Kelly Dobson, Lynn Festa, Matthew Fuller, Liza Hebert, Kara Keeling, Susan McNeil, Tara McPherson, Lisa Nakamura, Lisa Parks, Karen Tongson, and Nicholas Mirzoeff. I am indebted to many incredible student assistants who have helped shape this book: Lakshmi Padmanabhan, Ann-Kathrin Wagner, Daniela Wenzlaff, and Thomas Pringle. To the fantastic editorial team at MIT—Doug Sery, Matthew Abbate, Paula Woolley, Susan Buckley, and Susan Clark—I owe an enormous thanks.

To those who have been there for me for a very long time, I owe an inexpressible gratitude and unpayable debt: my friends Martha Fieltsch, Jennifer LeCouter, and Natalka Migus, who have helped to appreciate "nature's fireworks"; my family, Yeong Shik and Soon Jom Chun, Jeannie, Matthew and David An; and lastly my sweetie Paul Moorcroft, who has made this and so much else possible. Every day is a joy because of you, and you have helped me do things that I have never dreamed of attempting—thank you, (I think ;-)).

Fragments of this book have been published in *differences, Amerikastudien / American Studies, Theory Culture & Society, Ubiquitous Computing, Complexity and Culture,* and *Transforming Participation: Youth, New Media, Politics.* Part of chapter 4, published as "Habits of Leaking: Of Sluts and Network Cards," *differences* 26, no. 2 (2015): 1–28, was co-written with the wonderful Sarah Friedland.

Introduction: Habitual New Media, or Updating to Remain (Close to) the Same

New media exist at the bleeding edge of obsolescence. They are exciting when they are demonstrated, boring by the time they arrive. Even if a product does what it promises, it disappoints. If an analysis is interesting and definitive, it is too late: by the time we understand something, it has already disappeared or changed. We are forever trying to catch up, updating to remain (close to) the same; bored, overwhelmed, and anxious all at once. In response to this rapidly cycling and possibly manic depressive time scale, much analytic, creative, and commercial effort has concentrated on anticipating or creating the future, the next big thing: from algorithms that sift through vast amounts of data in order to suggest or predict future purchases to scholarly analyses that assess the impact of technologies that do not yet exist. What matters most: figuring out what will spread and who will spread it fastest.

But is this really the best approach? What does this constant move to the future, which dismisses the present as already past, erase? What do we miss if we assume new media are simply viral or disruptive? *Habitual New Media* counters this trend to analyze the present through soothsaying by revealing that **our media matter most when they seem not to matter at all**, that is, when they have moved from the new to the habitual. Search engines are hardly new or exciting, but they have become the default mode of knowledge acquisition. Smart phones no longer amaze, but they increasingly structure and monitor the lives of their so-called owners. Further, sites that have long since disappeared or which 'we' think have, such as *Friendster. com* (as of 2015, it was mainly a South Asian gaming site), live on in our clicks and our habitual actions, such as 'friending.' Whether or not a virus spreads depends on habits, from the regular washing of hands to practicing safe sex. **Through habits users become their machines**: they stream, update, capture, upload, share, grind, link, verify, map, save, trash, and troll. Repetition breeds expertise, even as it breeds boredom.[1]

At the same time, even as users become habituated to and become inhab-itants of new media, new media, as forms of accelerated capitalism, seek to undermine the habits they must establish in order to succeed in order to succeed. Habituation dulls us to the new; because of the shelter—the habitat—offered by habituation, the new is barely noticed. Most new com-modities are ignored, which is why grocery stores routinely change their layout to call attention to new products: the best way to break an estab-lished habit is to change the environment.[2] Further, new media cannot survive if they simply are disruptive—new, singular, for the first time—for what matters is not first contact (the mythic patient Zero 'responsible' for a viral outbreak) but the many ones that follow. Most concisely, **habituation and the new are the dreams and nightmares of new media companies.** New media remain in habits and in off-shored trash heaps.

New media, if they are new, are new as in renovated, once again, but on steroids, for they are constantly asking/needing to be refreshed. They are new to the extent that they are updated. (In this sense, hackers are software companies' best friends: users regularly download the latest corporate spy-ware in order to avoid potential security flaws.) **New media live and die by the update:** the end of the update, the end of the object. Things no longer updated are things no longer used, useable, or cared for, even though updates often 'save' things by literally destroying—that is, writing over—the things they resuscitate. (In order to remain, nothing remains, so now nothing remains even as everything does.) Things and people not updating are things and people lost or in distress, for users have become **creatures of the update. To be is to be updated:** to update and to be subjected to the update. The update is central to disrupting and establishing context and habituation, to creating new habits of dependency. To put it in a formula: **Habit + Crisis = Update.**

This book investigates the twinning of habits and crisis that structure networked time. It argues that if "networks" have become the dominant concept, deployed to explain everything new about this current era from social to military formations, from global capital to local resistance, it is because of what they are imaged and imagined to do. As I explain in more detail in chapter 1, "networks" render the seemingly complex and unmap-pable world of globalization trackable and comprehensible by transforming time-based interactions and intervals into spatial representations: they spa-tialize temporal durations and repetitions. Networks embody "glocal" com-binations by condensing complex clouds of interactions into definite, traceable lines of connection (or connections imagined to be so) between individual nodes across disparate locales. Network maps mediate between the local and the global, the detail and the overview. Their resolution 'illustrates' the relationship between two vastly different scales that have hitherto remained separate—the local and the global, the molecular and the molar—by reducing the world to substitutable nodes and edges. Not everything, however, becomes an edge. Imaged and imagined connections, this book reveals, are most often habits: things potentially or frequently repeated. Habit is information: it forms and connects. **Habits are creative anticipations based on past repetitions that make network maps the historical future.** Through habits, networks are scaled, for individual tics become indications of collective inclinations. **Through the analytic of habits, individual actions coalesce bodies into a monstrously connected chimera.**

Habitual repetition, however, as I explain in chapter 2, is also constantly undone by the other temporality of networks: crisis. As many others have argued, neoliberalism thrives on crises: it makes crises ordinary.[3] It creates super-empowered subjects called on to make decisive decisions, to intervene, to turn things around. Crises are central to habit change: in the words of Milton Friedman, creator of crises par excellence, "only a crisis—actual or perceived—produces real change."[4] Crises become ordinary, however, thwart change and make the present, as Lauren Berlant has put it, an impasse, an affectively intense cul-de-sac.[5] Crises make the present a series of updates in which we race to stay close to the same and in which information spreads not like a powerful, overwhelming virus, but rather like a long, undead thin chain.[6] Information is not Ebola, but instead the common cold.

Habit + Crisis = Update also makes clear the ways in which networks do not produce an imagined and anonymous 'we' (they are not, to use Benedict Anderson's term, "imagined communities") but rather, a relentlessly pointed yet empty, singular yet plural YOU.[7] Instead of depending on mass communal activities, such as reading the morning newspaper, to create national citizens, networks rely on asynchronous yet pressing actions to create interconnected users. In network time, things flow noncontinuously. The NOW constantly punctures time, as the new quickly becomes old, and the old becomes forwarded once more as new(ish). **New media are N(YOU) media**; new media are a function of YOU. New media relentlessly emphasize you: *Youtube.com*; What's on your mind?; You are the Person of the Year (see figures 0c.1 and 0c.2).

Networks trace unvisualizable interactions as spatial flows, from global capital to environmental risks, from predation to affects, by offering a resolution that pierces through the 'mass' or community to capture individual and preindividual relations.[8] Importantly, YOU is a particularly shifty

Figure 0c.1
Time magazine cover: *YOU as Person of the Year*. Reprinted with permission.

shifter: YOU is both singular and plural; in its plural mode, however, it still addresses individuals as individuals. This YOU, as I argue in chapter 3, is central to the changing value of the Internet, to the transformation of the Internet into a series of poorly gated communities that generate YOUs value. This YOU seeks to contain the leaky network that is the Internet. It perverts the Internet's wonderful creepiness through a logic that freezes memory into storage.

Further, Habit + Crisis = Update reveals the extent to which habit is no longer habit. Constantly disturbed, habit, which is undergoing a revival within the humanities, social and biological sciences, and popular literature, has become addiction. Habit has moved from *habes* (to have) to *addictio* (to lose—to be forfeited to one's creditor). Habit is now a form of dependency, a condition of debt. As the book's conclusion asserts, to be

Figure 0c.2
Time magazine cover: *YOU as Person of the Year*

indebted is not necessarily bad, but indebtedness must be complemented by a politics of fore-giving: of giving in excess and in advance. *Updating to Remain the Same: Habitual New Media* thus ends with a call to embrace the ephemeral signals that are constantly touching and caressing us. It concludes by refuting the politics of memory as storage—a society in which I do not remember, YOUs do not forget—by outlining a different kind of exposure and writing that repeats not to update, but to inhabit the inhabitable.

Contradictory Habits

Habits are strange, contradictory things: they are human-made nature, or, more broadly, culture become (second) nature. They are practices acquired through time that are seemingly forgotten as they move from the voluntary to the involuntary, the conscious to the automatic. As they do so, they penetrate and define a person, a body, and a grouping of bodies. To outline some of the contradictions habits embody (and habits are all about embodiment): they are mechanical and creative; individual and collective; human and nonhuman; inside and outside; irrational and necessary.[9]

Habits are both inflexible and creative. According to psychologist Wendy Wood, habits are voluntary actions initially taken on to achieve a goal; these actions, however, soon become autonomous programs.[10] As "memory chunks," habits are inflexible. Whereas a goal, such as weight loss, may be satisfied in many ways (exercise, various forms of dieting, etc.), there is only one way to satisfy a habit: doing it (eating potato chips while watching TV). This possible divergence of habits from goals can make habits, Wood argues, "vestige[s] of past goal pursuit."[11] At the same time, the involuntary nature of habit—its seemingly mechanical repetition of the same—combined with its tendency to wander, makes creativity and rational thought possible. According to Elizabeth Grosz, habit in the work of Gilles Deleuze, Félix Ravaisson, and Henri Bergson "is regarded not as that which reduces the human to the order of the mechanical ... but rather as a fundamentally creative capacity that produces the possibility of stability in a universe in which change is fundamental. Habit is a way in which we can organize lived regularities, moments of cohesion and repetition, in a universe in which nothing truly repeats."[12] Habit enables stability, which in turn gives us the time and space needed to be truly creative, for without habit there could be no thinking, no creativity, and no freedom.[13] Further, habit, as a form of second nature, reveals the power of humans to create new structures and reactions in response to their environment; it is, as Catherine Malabou writes, a sign of human plasticity.[14] Habit, unlike instinct, is learned, cultivated; it is evidence of culture in the strongest sense of the word. Moreover, habit itself is creative. As David Hume has observed, habit can create fanciful relations between things by invoking false experiences.[15]

This creative accrual of habit is central to personality, to subjectivity (or the lack thereof), and to ideology. Maurice Merleau-Ponty has argued, "my own body ... is my basic habit";[16] Gilles Deleuze has similarly stated, "we are habits, nothing but habits—the habit of saying 'I.'"[17] Drawing from this work, Clare Carlisle contends that habits, defined as what one has, conceal the fundamental emptiness of the self.[18] A habit, of course, is also a literal covering, and the nun's habit reveals that, even as habit covers and fits an individual, it also connects bodies.[19] According to Gabriel Tarde, this cohesion happens at a level beneath rationality and consciousness: modern man is a somnambulist who is linked to others through habit.[20] This link has been traditionally formulated in terms of class formation, most clearly in Pierre Bourdieu's work on *habitus*. *Habitus*, as the "generative principle of regulated improvisations"[21] ensures seemingly spontaneous harmony between members of the same class: "if the practices of the members of the same group or class are more and better harmonized than the agents know or wish, it is because ... 'following only [his] own laws,' each 'nonetheless agrees with the other.'"[22] Bourdieu was influenced by William James's argument that habit prevents class warfare by keeping "different social strata from mixing."[23] Habit, James writes in a widely cited passage, "is ... the enormous fly-wheel of society, its most precious conservative agent. It alone is what keeps us all within the bounds of ordinance, and saves the children of fortune from the envious uprisings of the poor. ... It holds the miner in his darkness, and nails the countryman to his log-cabin and his lonely farm through all the months of snow."[24] In this sense, habit is ideology in action; indeed, it is ideology as action, as opposed to false consciousness. According to Slavoj Žižek, ideology is structured like a fetish: we know very well that X is not Y, but we continue to act as though it were. We are "fetishists in practice, not theory."[25] Žižek makes this point by quoting Blaise Pascal's famous instructions on how to become a believer: "You want to find faith and you do not know the road. ... [L]earn from those who were once bound like you and who now wager all they have. ... Follow the way by which they began. They behaved just as if they did believe, taking holy water, having masses said, and so on. That will make you believe quite naturally, and will make you more docile."[26] This formulation links individual faith to others' habits. To acquire a habit, one deliberately learns from others: habits are forms of slow training and imitation that lead to belief, or at least the appearance thereof. Bourdieu similarly emphasizes the importance of practice to *habitus*. *Habitus* reveals the limits of rationality and regulations, for *habitus* cannot be explained through 'rules.'[27] This move to habit to understand the 'irrational' is also prevalent in institutional economics, which uses habit to critique game theory's reliance on rational choice theory.[28] Increasingly, habit is the productive nonconscious.[29]

Habits link not only humans to other humans, but also humans to nonhumans and the environment. Classically, crystal formation is a habit. Contemporary neurological examinations of the basal ganglia connect this evolutionarily 'ancient' part of the human brain, which is central to habit formation, to other animals: the basal ganglia is allegedly the same in fish and humans.[30] Further, habits link people to the environment. As Heidi Cooley states in her analysis of mobile technology, mobile devices are designed to be responsive and spontaneous, to work at the level of manual habits.[31] According to John Dewey, habits are "things done *by* the environment by means of organic structures or acquired dispositions."[32] Wood, among many other psychologists, has argued that habits are provoked by environmental cues, hence the importance of environmental changes and context—"disruption"—to habit change.[33] At the same time, though, habits also habituate. They enable us to ignore new things; they dull us to sensation and the environment.[34]

Given the diverse understandings of habit outlined above, it is tempting to portray habit as deconstruction embodied. Habits reveal the creative in the mechanical, the machinic in the creative. They trouble the boundary between self and other; they embed society in bodily reactions; they move between the voluntary and the involuntary. At a time when deconstruction is allegedly dead—killed by the very return to vitalist philosophy that grounds this return to habit—its spirit lives on in habit as *pharmakon*.[35] Perhaps. But the return to habit and to nineteenth- and early-twentieth-century theories of habit within sociology, psychology, and philosophy should make us pause. Why is habit—largely disregarded by critics during the era of Taylorism, in which humans were perhaps perversely theorized as rational agents—resurging in the era of neoliberalism, in which we are all individuals; in which, as Margaret Thatcher declared, "there is no such thing as society"?[36] **Are habits what endure as society within collectives in which there is no society?** What ideology remains as in allegedly postideological, post-class-based networks?

To answer these questions and to understand their implications, this book attends to the ways that habit itself is changing and to the contradictions between the many different understandings of habit outlined above. Change, of course, is central to the very notion of habits. However difficult it is to change habits, habits are acquired: they are not 'hardwired' involuntary actions, such as breathing. This is especially clear in the popular literature on habit, which focuses on self- and corporate improvement. This literature which, according to Amazon's data on the most highlighted passages in Kindle texts, is one of the most popular types of books, studies habits in order to change them, to improve the self.[37] Charles Duhigg has most famously argued that habit is a loop (see figure 0c.3), initially provoked by a cue and a reward. However, once a body is habituated, the person anticipates the reward, so that craving drives the loop. This explanation of habit reveals that something very strange is happening cloaked within this apparent renewal of habit. **Habit is becoming addiction: to have is to lose.**

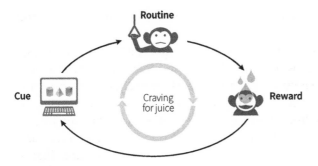

Figure 0c.3
The Habit Loop, from Duhigg, *The Power of Habit* (redrawn by Seungyeon Gabrielle Jung)

Examining the differences between nineteenth-century definitions of "habit," which are currently being rehabilitated (in particular Félix Ravaisson's, which is cited by almost every critical theorist), and Duhigg's loop makes this transformation clear. According to Ravaisson, habit is central to a living being's persistence in the face of material transformations to its body and environment: it is unity of diversity. Nonliving things thus do not have habits, because they are inanimate and uniform.[38] Further, habits are not instincts; they are not automatic reactions. Habit, rather, implies an interval necessary for creative response. It signals a change in disposition— a disposition toward change—in a being which itself does not change, even as it does. (My cells clearly change on a regular basis, but I am still me.) Habit frames change as persistence, as it habituates: it is a reaction to a change—to an outside sensation or action—that remains beyond that change within the organism. Through habit, we transform a change provoked by the outside into a change generated from the inside, so that receptivity is transformed into an unreflective spontaneity, beneath personality and consciousness. Habit therefore makes act and goal coincide, for habit occurs when an action is so free that it anticipates and escapes the will or consciousness (repetition breeds skill). The example Ravaisson offers of the moral person encapsulates this nicely. At first, becoming a moral person requires effort because there is a difference between the state of the person who wants to be moral and the state of morality. Gradually, though with effort, morality becomes effortless, and certain actions become pleasurable in and of themselves, so that, Ravaisson argues, pity for a certain person at a particular time becomes charity in general. Through habit, we become independent of both cue and reward, spontaneously producing actions and sensations that satiate and satisfy.

Ravaisson's definition of habit could not be more different than Duhigg's loop, in which habit is anything but spontaneous and self-sustaining. Rather, habit is driven by a craving that relies on a reward and a cue: it is addiction and dependence, not receptivity become spontaneity. Tellingly, Duhigg offers alcoholism as one of his key examples of habit, and he also draws heavily from examples in advertising. This transformation of habit into addiction is accompanied by a pathologization of habituation, that is, habit as *habes*. To be satisfied with what one has—to not want the next thing, the next upgrade—is to be out of sync. To hold onto what one has makes one a borderline 'hoarder,' in need of an intervention. Autonomous habituation supposedly disconnects. In contrast, 'habit as loss' habituates us to constant change, to the constant updating of habits needed to develop new dependencies. These dependencies drive 'networked society' and its logic of capture, crisis, and optimization. As Steven Shaviro has pointed out, networks thrive on debt.[39]

Neoliberal Habits: Privacy Depraved and Deprived

Habits are central to understanding neoliberalism, in particular to comprehending its simultaneous dissemination and contraction of privacy. Neoliberalism, to repeat a cliché, destroys the public by fostering the private. It leads to the rampant privatization of all public services, and, at least in the United States, gives private corporations the rights of citizens. Neoliberalism, as David Harvey succinctly explains, is "a theory of political economic practices that proposes that human well-being can best be advanced by liberating individual entrepreneurial freedoms and skills within an institutional framework characterized by strong private property rights, free markets, and free trade."[40] In a neoliberal society, the logic of the market has become its ethics; all human interactions, from love to education, become economic transactions to be analyzed in terms of costs and benefits. As Margaret Thatcher put it, neoliberalism "change[s] the soul"[41] by becoming, Michel Foucault argues, the "grid of intelligibility" for everything.[42] According to Wendy Brown, neoliberalism destroys democracy and the political by reconfiguring and constructing "all aspects of existence in economic terms."[43]

Neoliberalism emphasizes individual empowerment and volunteerism. In neoliberalism, the *homo oeconomicus*—the figure of the individual that lies at the base of neoliberalism—moves from "the [liberal] man of exchange or man the consumer" to "the man of enterprise and consumption."[44] Harvey asserts that neoliberalism does so by creating a general "culture of consent," in which the majority of people endorse neoliberal policies that actually hurt them, since these policies foster ever widening income disparities between the 99% and the 1%.[45] Neoliberalism's portrayal of itself as liberating drives this consent, for it has incorporated progressive 1960s (habits of) discontent with government while at the same time disassociating these protests from any critique of capitalism and corporations. Milton Friedman most forcefully linked economic freedom to political freedom: neoliberalism, he contended, is "a system of economic freedom and a necessary condition for political freedom" (although late in life he questioned the necessity of political freedom).[46] This economic and political freedom depends on information, for neoliberalism rests on the "proposition that both parties to an economic transaction benefit from it, *provided the transaction is bi-laterally voluntary and informed.*"[47] It is a system, as Brown notes, based on competition rather than equality or exchange.[48] In this 'open' system, individuals' habits—their abilities to quickly use freely available information—allegedly separate the winners from the losers.[49]

Neoliberalism's emphases on individual interest and market transactions spread the private (as market) and by doing so apparently destroy the private (as the intimate, darkened space necessary for growth and freedom); yet privacy traditionally was considered a state of deprivation: in monarchial systems, 'private' subjects, unlike public ones, could hold no power. More recently, as Oskar Negt and Alexander Kluge have argued in relation to 1970s Germany, the public sphere established its authority by dismissing the proletariat context of living as "'incomprehensible' ... ultimately, it becomes a private experience."[50] This bourgeois denigration of the private, however, also coincided with the reification of the bourgeois private or intimate as a shield from publicity (a shield pointedly not offered to [former] slaves). John Stuart Mill most famously maintained that liberty depends on the separation of public and private spheres, effectively transforming privacy from something privative to something sacred.[51] Jürgen Habermas and Hannah Arendt, among others, have argued that the cultivation of a refined and refining private sphere (or more properly an intimate, family sphere) was central to the emergence of rational public discourse.[52] To all these thinkers, from Mill to Kluge, the domestic was key to defining the boundaries between public and private, for the walls of the home sealed the private from the public. As Negt and Kluge put it, "the family ... [both] protect[s life] from the world of work and ... [erects] the libidinal structure ... whereby individuals become capable of being exploited"; more positively, Arendt states, "these four walls, within which people's private family life is lived, constitute a shield against the world and specifically against the public aspect of the world."[53] The walls of the home, however, are no longer secure, if they ever were: there is no shield from competition, for the twin forces of media and market compromise domestic 'protection.' **Privatization is destroying the private, while also fostering state surveillance and security as house arrest.**

Clearly, this spread and wane of the private extends far beyond social media and the Internet. The Internet, however—with all the hype portraying it as ideal public sphere, ultimate surveillance machine, hotbed of radicalism, and/or dangerous pornographic badlands—remains central to understanding how this depraved and deprived privacy is negotiated and implemented. New media call into question the separation between publicity and privacy at various levels: from technical protocols to the Internet's emergence as a privately owned public medium, from *Google.com*'s privatization of surveillance to social networking's redefinition of "friends." As I explain in chapter 3, social media are driven by a profound confusion of the private and public. The very notion of a "friend," initially viewed as a way to restrict communications in social media sites such as *Friendster.com*, has led to various 'crises' that stem from breaches in the boundaries between private and public, friend/boss/mother. Tellingly, Senator Dan Coats, during the U.S. Senate debate on the Communications Decency Act—the first attempt by the U.S. government to regulate content on the Internet, an act lodged within a larger one that also privatized the Internet backbone—described the Internet as "taking a porn shop and putting it in the bedroom of your children."[54] The Internet has been perceived as an unstoppable window that threatens to overwhelm the home and existing zoning laws. As the second half of this book makes clear, **Internet users are curiously inside out—they are framed as private subjects exposed in public.** They are children with porn shops in their bedroom.

Rather than accepting the current terms of the debate, which reduce privacy to secrecy or corporate security and which blame users for systemic vulnerabilities, *Updating to Remain the Same* asks: What happens if we take seriously the leakiness of new media? As I elaborate further in chapter 1, our networks work by promiscuously exchanging information: a networked personal computer is an oxymoron. Our networks leak, however, not only at the level of technical infrastructure, but also at the level of content. New media erode the distinction between the revolutionary and the conventional, public and private, work and leisure, fascinating and boring, hype and reality, amateur and professional, democracy and trolling. As Ethan

Zuckerman has argued, the same medium used to organize political protests disseminates countless images of cute cats: banality of content supposedly equals stability of platform.[55] This combination of gossip with politics is not an unfortunate aspect of new media and digital culture, but the point. New media blur these distinctions because they are part of the postindustrial/ neoliberal economy.[56] This blurring drives the logic of crisis allegedly provoked by these leaks (chapter 2), as well as the massive extension and contraction of privacy and of the state that is neoliberalism (chapters 3 and 4). This blurring reveals a logic of containment, which is always imagined as already transgressed.

To state the obvious, this notion of new media as erosion depends on the prior acceptance of networks as fundamentally personal and private, a notion that depends on certain habits of privacy that often undermine the very privacy sought.[57] Further, as the second half of this book elaborates, this logic thrives via an epistemology of outing, which constantly exposes open secrets. To break from this logic of leaking and outing, this book contends that we need to embrace the fundamentally nonpersonal nature of our networked communications. To be clear, I am not arguing that state and corporate surveillance are irrelevant or that the leaks exposed by Edward Snowden amongst others should be ignored. What I am saying is that what is most surprising and alarming about the Snowden revelations is the fact that they counted as revelations. Not only have whistle-blowers been revealing the existence of deliberately constructed back doors in email programs and massive packet sniffing on Internet backbones; promiscuous networks operate by making users vulnerable.[58] Rather than accepting a 'security' which renders all public transmissions (many without which there could be no communications) into 'leaks' and thus fosters a fundamental ignorance about networking technologies, chapter 4 reveals that we need to fight for the right to be vulnerable and not attacked. That is, we need modes of networked inhabitation that engage with and buttress publicity, rather than seek a false refuge in privacy.

Habits Not Viruses

To foreshadow these modes of networked inhabitation with which I end this book, I want briefly to outline how habits refigure what we too quickly call affective contagion or disruption. In the era of network maps, which trace node-to-node contact, everything, from obesity to marketing, supposedly spreads virally. While clearly some things do spread in this manner, not everything that mimics node-to-node infection is a virus. This fact was made clear in responses to the controversial study that 'revealed' that obesity spreads person to person.[59] In this study, the authors reused and correlated data from the Framingham Offspring study, initially set up to diagnose risk factors for heart disease, to create a social map for this group.[60] Analyzing this data, they found that "the risk of obesity among alters who were connected to an obese ego (at one degree of separation) was about 45% higher in the observed network than in a random network. The risk of obesity was also about 20% higher for alters' alters ... and about 10% higher for alters' alters' alters. ... By the fourth degree of separation, there was no excess relationship between an ego's obesity and the alter's obesity."[61] Based on these findings, they concluded that social proximity plays a more important role than physical closeness in the spread of obesity. Further, the strength of the friendship matters: "between mutual friends, the ego's risk of obesity increased by 171% ... if an alter became obese. In contrast, there was no statistically meaningful relationship when the friendship was perceived by the alter, but not the ego."[62]

There has been much controversy around this report, and for many it has become a poster child for bad network analysis.[63] The soundness of the statistical model used, the authors' guiding assumptions, and their conclusions have all been scrutinized. These critiques have argued that, even if these observations are correct—even if these correlations are real—the conclusion is not true. In particular, as Cosma Rohilla Shalizi and Andrew C. Thomas have shown, it is mathematically difficult to distinguish between viral spread and peer homophily, the tendency of peers to "behave similarly as a result of 'correlated effects' such as common environmental shocks or shared characteristics rather than social influence."[64] It is equally probable that the "spread" of obesity is due to the tendency of peers to hold and develop similar habits, and Shalizi and Thomas's definition of "homophily" corresponds almost perfectly to Bourdieu's notion of *habitus*—people who are "alike" spontaneously act the same—without, of course, Bourdieu's emphasis on class.

The vehemence of the responses against the report reveals the (still) close tie between medically accepted causality and physiology (obesity clearly is not a virus; it should not be confused with one), as well as a commitment to therapeutically useful analyses. Although the initial Framingham Heart survey was criticized for the narrowness of its cohort, it has been praised generally for helping to identify major cardiovascular disease risk factors, from smoking to cholesterol to heredity.[65] In contrast, the obesity study offers no clear therapeutic benefit. Even if the correlation is true, it is unclear how it could be useful in preventing or treating obesity, since the implied solution is the social isolation of obese people. For whom, besides insurance companies, is this correlation—the revelation regarding mutual habit formation—useful? These studies, as discussed in chapter 1, are not designed to foster justice.

A focus on habit moves us away from dramatic chartings and maps of "viral spread" toward questions of infrastructure and justice. To take the example of the 2014–2015 Ebola outbreak, a discussion of habit would move us away from what Priscilla Wald has called an "outbreak narrative"— a narrative that manages and diagnoses communicable disease by concentrating on identifying an emerging infection and the global networks through which it travels and is contained—toward understanding the conditions that made this spread possible: from crumbling medical infrastructures to new patterns of mobility brought about by globalization to the lingering impacts of colonialism and civil wars.[66] Studying habit moves us away from theories of affective spread and disruptions, not because they are not important, but rather because they leave us with questions: Why and how are so many things seemingly ignored? Why and how do things linger? Why and when does repeated contact produce habituation, instead of sensitization? As Lisa Blackman has argued, habit is linked to affect modulation; it enables both the regulation of and potential for engaging the new.[67] Habit moves us from the rapid time scale of viral infections and disruptions toward the slower and more stable time frame of homophily, a frame better suited to explain the 'undead' nature of information spread. Further, a focus on habit moves us away from an epistemology of outing, in which we are obsessed with 'discovering' 'Patient Zero,' as though knowing the first case could solve all subsequent problems. Relatedly, an emphasis on inhabitation and lingering turns attention away from the networked viral YOU to the possibility of a 'we.' Habit, with all its contradictions, is central to grasping the paradoxes of new media: **its enduring ephemerality, its visible invisibility, its exposing empowerment, its networked individuation, and its obsolescent ubiquity.**

YOU Are Here

The book is divided into two parts, "Imagined Networks, Glocal Connections" and "Privately Public: The Internet's Perverse Subjects." The first part unpacks how and why 'the network' has become the defining concept of our era. Revealing that networks have become key because they are imagined as ending postmodern confusion, these chapters trace how networks make possible groupings based on individual and connectable YOUs. They also elaborate on crisis as structuring the temporality of networks. The second part of the book develops more fully the inversion of privacy and publicity that drives neoliberalism and networks. Highlighting how networks capture subjects through users 'like YOU'—that is, users who like YOU ('friends') and those determined to be like YOU ('neighbors')—these chapters both document the epistemology of outing that drives this logic and outline ways of inhabiting this outing. This part therefore concludes by examining how survivors of abuse embrace templates and repetition in order to inhabit hostile networks. Through habit—the scar or memory of others that lives on in the self—users can inhabit alterity.

Further, each chapter of this book is framed around one of the following aphorisms of habitual new media:

1. Always Searching, Never Finding
2. Habit + Crisis = Update
3. The Friend of My Friend Is My Enemy (and Thus My Friend)
4. I Never Remember; YOUs Never Forget.

These aphorisms highlight both the dilemmas and opportunities opened by N(YOU) media: the dilemmas and opportunities YOU face as a small *s* sovereign, but also the dilemmas of and opportunities for shifting the YOU, for keeping this shifter shifty. Each chapter revisits the question of habits through critical revisitings of habit. Rather than return to the alleged primary source, each chapter seeks to understand the current resurgence of habit by exploring how habits resuscitate certain critical thinkers. In other words, each chapter investigates how both new media and critical theory remain. Each chapter is also prefaced by a section, structured like a comment in a programming language, that explores an example addressing the stakes of that chapter.

Chapter 1, Habitual Connections, or Network Maps: Belatedly Too Early (Always Searching, Never Finding), outlines the relationship between habitual new media and networks. Responding to the questions "Why have networks become a universal concept employed by disciplines from sociology to biology, media studies to economics?" and "How has 'It's a network' become a valid answer, the end, rather than the beginning, of an explanation?," this chapter argues that networks have been central to the emergence, management, and imaginary of neoliberalism—in particular to its conception of individuals as collectively dissolving society. Tracing the ways in which networks, or more precisely mappings of networks, were embraced as a way to dispel the postmodern confusion that dominated the late seventies and early eighties, the chapter demonstrates that networks allow us to trace if not see—that is, to spatialize—unvisualizable flows, from global capital to environmental risks, from predation to affects. Although they enable a form of cognitive mapping that links the local to the global, networks produce new dilemmas: neoliberal subjects are now forever mapping, but more precarious than ever; they are forever searching, but never finding. Further, networks are belatedly too early. They are both projections and histories; they are both theory and empirically existing entities. To begin to imagine networks differently, this chapter thinks about new media in terms of habitual repetition and constitutive leaks. Information, it argues, is habit.

Chapter 2, Crisis, Crisis, Crisis, or The Temporality of Networks (Habit + Crisis = Update), addresses how networks make crises habitual. It argues that crisis is new media's critical difference: its norm and its exception. New media thrive on crises, which are now both everyday occurrences and extraordinary occasions. Crises differentiate and interrupt the constant stream of information, marking the temporarily invaluable from the mundane and offering users tastes of real-time responsibility. They also threaten to undermine this agency by catching and exhausting users in a never-ending series of responses. This chapter places crises in the context of automated habits and codes (indeed, habit as code), which have been designed supposedly for our safety, and it reveals that crises and codes are complementary because they are both central to the emergence of what appears to be their antithesis: user empowerment and agency. Codes and crises together produce (the illusion of) mythical and mystical sovereign subjects, who weld together word with action, norm with reality. To combat this and to exhaust exhaustion, this chapter ends by emphasizing the plasticity of habits and memory.

Chapter 3, The Leakiness of Friends, or Think Different Like Me (The Friend of My Friend Is My Enemy (and Thus My Friend)), investigates the odd transformation of the default Internet user from the lurker to the friend as indicative of a larger encroachment and recession of the private. It contends that a banal and impoverished notion of friendship grounds the transformation of the imagined Internet from the thrillingly dangerous and utopian "cyberspace" to a friendlier, more 'trustworthy' Web 2.0 (3.0, 4.0, etc.). This desire for a reciprocal and authenticating, if not entirely authentic, type of intimacy—for friendship as 'friending'—echoes and buttresses the notion of users as neoliberal, small s sovereigns. This mode of empowerment, however, threatens to turn the Internet into a series of gated communities, of enclosed open spaces dominated by what I call YOUs value. Further, these attempts to make the Internet more familiar and reciprocal have made it a more nasty space, in which users are most in danger when they think they are most safe, and in which users place others at risk through their sometimes genuinely caring actions. This chapter, though, does not simply indict networks and desires for intimacy and contact. Rather, it and the next chapter point to other models of friendship, which emphasize unreciprocal or unreciprocating relations, in order to theorize the possibilities for democracy that stem from networked vulnerabilities.

Chapter 4, Inhabiting Writing: Against the Epistemology of Outing (I Never Remember; YOUs Never Forget) draws the book to a close by exploring alternative ways to inhabit networks. It starts with the observation that network safety is increasingly framed in gendered terms: the need to protect teenage girls drives calls for online transparency; at the same time, the 'bad user' is the 'girl' who threatens her own (and other people's) safety through her promiscuity. Tracing this logic in several high-profile bullying cases and linking it to the long history within the United States of defining the right to privacy in relation to white, bourgeois femininity, this chapter argues that, rather than accept this safety (which is in fact no safety), we need to fight for the right to loiter in public. Focusing in particular on Canadian teenager Amanda Todd's moving video, posted months before her suicide, in which she uses note cards to relay her story and shelter her face (a technique used by many abused young adults and by *wearethe99percent.com*), the chapter argues for a shadowy inhabitation of networks based on a right to be vulnerable and not attacked. Rather than a politics based on vengeance, we need a politics of fore-giveness and remembering.

These sections can be read out of order: those most interested in social media might want to start with part II, "Privately Public: The Internet's Perverse Subjects."

New Media: Wonderfully Creepy

To conclude this introduction, I want to place this book in the context of the two others that preceded it—*Control and Freedom: Power and Paranoia in the Age of Fiber Optics* (2006) and *Programmed Visions: Software and Memory* (2011)—for *Updating to Remain the Same: Habitual New Media* completes my trilogy on how computers emerged as a form of mass media to end mass media by replacing the mass with the new, the 'we' with the YOU. This replacement encapsulates the promise, and threat, of new media. If mass media produced consistent forms to create consistent, coherent audiences, new media thrive on differences to create predictable individuals.

The first book responded to the question: How did a control technology become bought and sold as a technology of freedom? To answer this question, it began by tracing how the Internet, a technology that had existed for decades, became "new media" when it became fictional: when it was portrayed as embodying cyberspace, the long lost Athenian public sphere, and all sorts of other theories and dreams. This remarkable transformation of a control protocol into a medium of freedom entailed another change: the reduction of freedom to control. Against this limiting view—in which one is free only when one is in control—*Control and Freedom* argued that freedom always exceeds control: freedom makes control possible, necessary, and yet never enough.

The second book started with question: How and why have computers become key metaphors and tools for understanding and negotiating our increasingly complex world? In response, it revealed the centrality of computers—understood as networked hardware/software machines—to neoliberalism: they are essential to organizing and managing, assessing and predicting—that is, programming—populations and individuals. In making this argument, *Programmed Visions* focused not simply or primarily on the varied tasks computers now accomplish, but rather on the very notion of programmability—a vision of rationality that covers over incomprehensible and opaque executions—that they have come to encapsulate. It did so not in order to condemn and move beyond computer interfaces and software, but rather to understand how this combination of visibility with invisibility, of past experiences with future expectation, makes new media such a powerful thing for each and all.

Both *Control and Freedom* and *Programmed Visions* emphasized the generative power of the paradoxes each started from, for these contradictions ground the appeal of new media and the centrality of new media to neoliberalism.

Updating to Remain the Same continues these investigations into the founding paradoxes of new media and seeks to theorize the possibilities for inhabiting networked vulnerabilities. Like the books before it, it suggests that, if our desires for something like secure intimacy and for computers as 'personal' put us at risk, these dangers can be best attenuated not through better or more security, but rather through a wary embrace of the vulnerability that is networking and through a reimagination of networks. Networks are not maps. The constant habitual actions invoked by and provoked for connections reveal that networks, if they are anything, are everything outside of neat lines and edges. To engage this everything, we need to forego the desire to reduce memory to storage, and we must develop a politics of fore-giving that realizes that to delete is not to forget, but to make possible other (less consensually hallucinatory) ways to remember.

Interlude: THEY→YOU

Once upon a time, in the era of mass media, the world was divided into 'we's and 'they's. The 'We' and 'they' were odd subjects, and, although they seemed opposed, they were really mirror images of each other. They were mass subjects: dense aggregates that were, or could be treated as, uniform, as the same throughout.

Martin Heidegger most (in)famously outlined the 'they' subject in *Being and Time*. The 'they' was *Dasein* in its everyday existence, entangled in the world. Even though the I was essentially the 'they,' 'they' was called 'they' in order to make the I feel singular and individual, "to cover over one's essential belonging to them."[1] To belong was to be average, indistinguishable from all the others entangled in the 'they.' Even worse, as one immersed oneself in the 'they,' not only did others become less explicit and discernible, the 'they' became more dictatorial: "We enjoy ourselves and have fun the way *they* enjoy themselves. We read, see, and judge literature and art the way *they* see and judge. But we also withdraw from the 'great mass' the way *they* withdraw, we find 'shocking' what *they* find shocking."[2] As this statement makes clear, the 'they' embodies public engagement: a being in public, entangled with others. To emerge as an independent (private) I, *Dasein* had to engage in an authentic struggle; but *Dasein* was entangled initially in the 'they.'

The 'they' ensnared *Dasein* in three ways: through idle talk, curiosity, and ambiguity. The 'they' communicate through gossiping, rather than authentic understanding, for the act of speaking matters more than the content of speech. 'They' do not comprehend or verify, but instead simply pass the word along. This groundless chatter makes things public: "Idle talk is the possibility of understanding everything without any previous appropriation of the matter. ... Idle talk, which everyone can snatch up, not only divests us of the task of genuine understanding, but develops an indifferent intelligibility for which nothing is closed off any longer."[3] Curiosity, which

Heidegger defines as a distracted "not-staying," similarly undermines com-
prehension and "seeks novelty only to leap from it again to another novelty. ... [It seeks] restlessness and excitement from continual novelty and
changing encounters."[4] Idle talk and curiosity, combined with ambiguity of
interpretation, ensure that the truly new is outdated as soon as it emerges.
They also make action seem inconsequential; because 'they' are obsessed
with interpreting and guessing things in advance, "carrying things out and
taking action [becomes] something subsequent and of no importance."[5] To
know is to not have to do.

To some, the subject of social media is the 'they' on speed. What is
Twitter.com—in particular, retweeting—if not idle talk? *Twitter.com* in its
very name shamelessly embraces inconsequential chatter: traditionally,
"twitter" referred to the chirping of birds and girls. (Daniel Paul Schreber's
world was filled with twitter: birds that constantly asked him, "Are you not
ashamed?")[6] Social media are dominated by curiosity: a drive to know—to
guess—that moves quickly to the next new intrigue. Everything is so five
minutes ago.

Whether or not one subscribes to such a harsh view, there is a crucial
difference between the mass 'they' and the subject of social media. Rather
than dissolving into the 'they,' the subject of social media always remains
distinguishable (i.e., trackable in its transmission of gossip). Again, what
emerges is a YOU: YouTube, YOU as *Time*'s person of the year. This YOU is
relentless and would seem to mask the fact that YOU are 'they.' The solution: a Heideggerian struggle for the (dis-)solution of the 'they' through the
I, through the realization of the singularity of one's death?

No, for let's not forget what unfolded after that original call. *Dasein* as
always already guilty indeed.

To chart a different path and to underscore the difference between 'they'
and YOU, let us start with Jean Baudrillard at his most perverse: "The
Masses: The Implosion of the Social in the Media." According to Baudrillard, most assessments of mass media assume that the silence of the masses
is a problem to be fixed. One thus needs to empower the masses by giving
them a voice, an authentic way to respond. (This is difficult because the
masses are also profoundly affected by all attempts to measure them, such
as polls.) Rather than endorse attempts at empowerment through participation, Baudrillard hypothesizes that this silence—this indifference to
politics—is ironic and antagonistic. It is "an original strategy, an original
response in the form of a challenge."[7] It is a strategic disappearance—the
masses are a black hole that absorbs rather than radiates—that reacts to the

demand to be liberated, rational, and authentic subjects by refusing choice and subjectivity altogether. Silence is the revenge of the object.

Regardless of whether Baudrillard's analysis is correct (his conflation of noisy indifference with silence is particularly suspect, for it is unclear how the noisy and collective watching of a football game counts as silence), an ironic, taunting silence of the masses is impossible online. Whether or not YOU respond, YOU constantly register and are registered—YOUR actions are captured and YOUR silence is made statistically significant through the actions of others 'like YOU.' (Again, YOU as both singular and plural.) YOU register through YOUR habits: habits distribute unity within diversity. **The media have imploded in the social. YOU are a character in a drama called Big Data.** YOU are knitted into a monstrous chimera because users perversely follow the mantra "If you see something, say something." (YOU are constantly reporting what YOU do not see.)

So, how do we understand this YOU, in whom users are captured, recognized, and updated? More importantly: How can users inhabit this singular plural YOU, precisely through those repetitions that Heidegger found so insipid? What future lies in idleness?

Part I   Imagined Networks, Glocal Connections

'The network' has become a defining concept of our epoch. From high-speed financial networks that erode national sovereignty to social media sites like *Facebook.com* that have transformed the meaning and function of the word 'friend,' from *Twitter.com* feeds that foster new political alliances to unprecedented globe-spanning viral vectors that threaten worldwide catastrophe, networks seem to capture everything that is new or different about our social institutions, global formations, and political, financial, and military organizations. From Bruno Latour's discussion of Actor Network Theory (ANT) to the discipline of network science, from Jean François Lyotard's evocative description of the postmodern self as a "nodal point" to Tiziana Terranova's analysis of global network culture, 'networks' are also central to theorizing networks.[1]

Why? What is it about 'the network' that makes it such a compelling and universal concept, deployed by disciplines from sociology to biology, media studies to economics? Why do we believe the network (however described) to encapsulate our "current social formation?"[2] How and why has "It's a network" become a valid answer—the end, rather than the beginning, of an explanation?

'Network' is an odd, almost contagious concept. Networks are both actually existing realities and theoretical abstractions. They are both planning diagrams and their result, both description and elucidation—they are theoretical in all senses of the word. Networks not only compromise the distinction between illustration and explanation, they also erode the boundaries between the many disciplines that employ networks, from economics to literary studies, from political science to biology. The study of networks thus oddly mirrors its subject, for the examination of networks leads to the formation of ever more networks, making it difficult to separate network analyses from networks themselves.

The next two chapters reveal that the power of networks stems from how they are imaged and what they are imagined to do. In particular, they are central to the ascendency, management, and imaginary of neoliberalism; their conversion of the world into nodes and edges—agents and connections—is imagined as dissolving postmodern confusion. They figure what was once considered unimaginable: global capitalism, environmental risks, nebulous affects, and "unresponsive" masses. Networks also encompass and interconnect the technological and the social. Across all diverse understandings of them—from the scholarly to the popular— networks are social technologies. They enable communication between people and buttress social apparatuses, allegedly making the social a technological product (hence the term "twitter revolutions"). They also make technologies social: currents flow in electrical networks from site to site.

To be clear, by focusing on networks as imagined, I am not arguing that networks are fanciful objects that do not exist.[3] Rather, I am both extending and revising Benedict Anderson's influential but much critiqued formulation in *Imagined Communities: Reflections on the Origin and Spread of Nationalism* that nations are "imagined political communit[ies]."[4] Anderson argues that nations are imagined "because the members of even the smallest nation will never know most of their fellow-members, meet them, or even hear of them, yet in the minds of each lives the image of their communion."[5] They are communities because, regardless of actual disparities, they are imagined as a "horizontal comradeship."[6] In making this argument, Anderson stresses the importance of print capitalism, such as the circulation of novels and newspapers, which makes time seem homogeneous and empty. Novels create an imagined community by intertwining the interior time of the novel's characters with the exterior time of the reader, so that together they become "a sociological organism moving calendrically through homogeneous empty time."[7] Novels, in other words, create a time and space in which readers and characters become a 'we/they': the novel's typical third-person narrative creates a general sense of nearness, rather than a targeted and individuating connection.[8] Newspapers, through their regularly planned obsolescence, create daily "extraordinary mass ceremonies," so that, reading a newspaper in the privacy of her own home, "each communicant is well aware that the ceremony he performs is being replicated simultaneously by thousands (or millions) of others of whose existence he is confident, yet of whose identity he has not the slightest notion."[9] Imagined synchronous mass actions create an imagined community in which the multiple 'I's are transformed every morning into a 'we' that moves together through time (hence *KONY 2012*'s invocation of the *New York Times*, discussed shortly).

Newspapers, however, are not what they used to be. As the early-twenty-first-century crisis in print publications has made clear, 'we' can no longer be certain of these extraordinary mass ceremonies, if 'we' ever could. From both perspectives—the intimate and the global—the notion of a national 'we' seems quaint at best, even as it persists in 'our' imagining of the global. This provokes the obvious question: Does this mean the death of imagined communities? Rather than participate in a long and endless debate over whether new media enable or destroy communities—whether they create new online communities or alienated, disconnected users; whether they create dangerous swarms or empowered networked individuals—this section focuses on new media as generating imagined networks. These networks generate a YOU rather than a 'we.' This YOU, produced through the act of mapping, is modulated temporally via asynchronous habits and crises, rather than mass ceremonies. Collective habits, such as reading the paper, still matter, but they matter and collect differently. As I elaborate in chapter 2, rather than enabling a "homogeneous empty time"[10]—a time that buttresses notions of steady progress—networks produce a series of crises or 'nows' that create bubbles in time. In these constantly updating bubbles, the new quickly ages and the old is constantly rediscovered as new. Hence, networks do not imagine a collective entity traveling together through time, but instead a series of individuals that (cor)respond in their own time to singular, yet connected, events.

Imagined Networks Are More and Less Than Communities

Imagined networks are glocal collectives that link the social-historical to the psychical, the collective to the individual. They are combinations that form definite, traceable lines of connection (or connections imagined to be so) between individuals across disparate locales. Network maps mediate between the local and the global, the detail and the overview—their resolution figures the relationship between two different scales: the individual and the typical, the molecular and the molar.[11] Even when these imagined networks coincide with the nation-state, they transform what is meant by 'nation' (and also 'race').[12] Networks are based on YOUs value: a series of interconnecting and connected YOUs. Since YOU in its plural mode still addresses individuals as individuals, networks are very different from communities, which create a new identity, a 'we', from what is held in common (even if, as Maurice Blanchot has argued, that "what" is incommensurability).[13] In a network, when a 'we' or mass simultaneous action happens, the network can crash: from multiple simultaneous hits on a website to synchronous and overwhelming electricity demand, from popular fakesters to flashmobs, the communal (both technical and nontechnical) can bring down networks. The 'we' is a temporary network weapon.

Always Searching, Never Finding

=begin *KONY 2012 (yes, really, KONY 2012 =P)*
Considered in 2012 to be "the most viral video of all time," *KONY 2012*, as many have noted, defies the usual formula for viral videos: there are no cute cats, no hallucinating children, and no bizarre acts of self-humiliation.[1] Instead, it is a thirty-minute video, ostensibly about a decades-long conflict in Uganda, that asks its viewers to help capture indicted war criminal Joseph Kony. The premise is simple: Joseph Kony is free because he is not famous; no one knows his name. The filmmaker Jason Russell therefore sought to make Kony a celebrity by recruiting a massive army of volunteers to contact influential Hollywood stars, businesspeople, and lawmakers. The assumption: at all levels, celebrity makes things happen.

Championed by Hollywood stars such as Oprah Winfrey, *KONY 2012* did make Kony—or at least *KONY 2012*—famous. This video reached 93 million views on *YouTube.com* within the first month, not only because of these endorsements, but also because of its dissection by pundits, policy makers, and academics, who mostly highlighted the obvious: the fact that *KONY 2012* had very little to do with Uganda. Although framed as a call to change Ugandan politics, it provided little to no historical or political context for Joseph Kony. At best, the video read as an updated (younger, more participatory, more female) version of "white man's burden"; at worst, it evoked the specter of the lynch mob.

Such critiques, though correct, miss the point. If anything, *KONY 2012* was about enacting and imagining the power of social networks.

The introductory frames of the video make this focus clear. The video begins with the words: "NOTHING IS MORE POWERFUL THAN AN IDEA WHOSE TIME HAS COME/IS NOW." It then

Figure 0f.1
Earth from space, *KONY 2012*

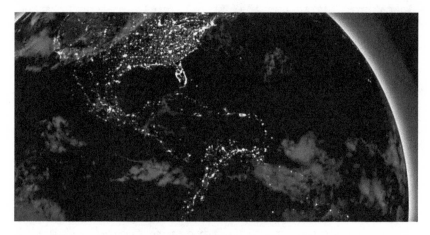

Figure 0f.2
The Dark Globe, focused on the United States of America, *KONY 2012*

moves (via some abstract frames signifying motion) from the iconic image of the earth from space (figure 0f.1) to an image of the darkened globe, featuring brightly lit U.S. urban hubs (figure 0f.2).

The voice-over narration then begins:
Right now, there are more people on Facebook than there were on the planet 200 years ago.
Humanity's greatest desire is to belong and connect, and now, we see each other; we hear each other,
[from intercut video: Grandpa I love you; I love you; why won't it take a picture?]
We share what we love, and it reminds us what we all have in common.
[Intercut video: Dug out alive and well after 7 and a 1/2 days;
If you believe in yourself you will know how to ride a bike! Rock and roll!;
Now, technically your device is on. Laughter. Can you tell? Crying. Oh that's exciting.]
And this connection is changing the way the world works. Governments are trying to keep up.
[Intercut video: Now we can taste the freedom; chanting.]
And older generations are concerned
[Intercut video: Many people are very concerned about tomorrow. They could get worse next year.]
The game has new rules.
[Visual: countdown begins, starting at 27:00f.]
The next 27 minutes are an experiment, but in order for it to work, you have to pay attention.
[Visual: Meteor]

Featuring viral and news videos and punctuated by repeated screen shots of users clicking "Share" (figure 0f.3), *KONY 2012* is a blatant and banal celebration of social networking sites that frames the imagined capture of Joseph Kony as proving the power of social networks: the goal is figured as a *New York Times* cover with the headline "KONY CAPTURED" (figure 0f.4).

Figure 0f.3
Repeated "Share" frame, *KONY 2012*

Figure 0f.4
Success: KONY CAPTURED, *KONY 2012*

This reference to the *New York Times* testifies to the
residual power of newspapers to imagine communities. It
also reveals the U.S.-centric, liberal nature of this
allegedly global "human" vision, as do the video's
illustrations of connection and globalization: the opening
segment includes pictures of racially diverse sets of

Figure 0f.5
From *KONY 2012*

Figure 0f.6
From *KONY 2012*

people, but these persons are carefully segregated into
separate frames, except in the image in which a Haitian
boy is saved by U.S. emergency workers (figures 0f.5,
0f.6, 0f.7, 0f.8). Serving as a link between these
segregated images, the disembodied male voice-over creates
a plural 'we' and narrates humanity as driven by a common
desire to connect.

Figure 0f.7
From *KONY 2012*

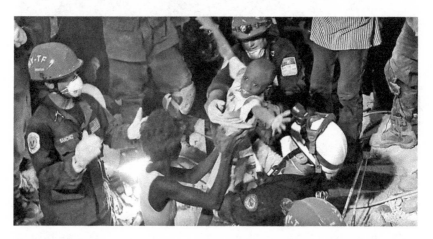

Figure 0f.8
From *KONY 2012*

This common desire to connect separates at another level, for it targets individuals as individuals, rather than as a group. Although it uses the pronoun "we" to link together individual stories, it refers to the viewer(s) relentlessly as YOU, in both the singular and plural senses of the word. In the section of the video in which Russell explains his mission, he states:

"We're going to stop them."
I made that promise to Jacob, not knowing what that would mean, but now I do.
Over the past nine years, I have fought to fulfill it, and the fight has led me here, to this movie you're watching, because that promise is not just about Jacob, or me, it's also about you.
And this year, 2012, is the year we can finally fulfill it, and if we succeed, we change the course of human history, but time is running out.

The "you" here is pictured as a yearbook spread in which blurry images of mainly teenage girls are placed next to each other: together, but identifiable and separate (figure 0f.9).

Figure 0f.9
YOU, from *KONY 2012*

The "you" is imagined as both singular and plural. It
directly addresses the viewer, but it also offers her an
image of herself in a plural form. Rhetorically, the "you"
here is an apostrophe: as Barbara Johnson has noted,
apostrophe "by means of the silvery voice of rhetoric …
calls up and animates the absent, the lost, and the
dead."[2] This "you," which you can only become through an
act of misrecognition (you aren't really these "yous"),
is also a demand. Should you (mistakenly) identify with
this absent position (and thus in turn animate the "I,"
the filmmaker), you must respond to what KONY 2012 frames
as a crisis. Time is running out; for some reason this
decades-long crisis must end in 2012. By accepting the
"you," a "we" emerges, but this "we" is deliberately
temporary: it is brought together for a specific task
now.[3] The "we" indicates a specific moment, a now, in
which "you" becomes a "we" at the moment of contact; in
contrast, the "you" is curiously permanent. It precedes
and remains after the "we" fades.[4]

KONY 2012 has been criticized as a form of
"slacktivism": online activism that reduces humanitarian
interventions to banal, ineffective acts, such as
contacting famous people in order to make Kony/KONY
famous. This critique elides the questions: What form of
humanitarianism is not a form of slacktivism? How
different is emailing celebrities such as Angelina Jolie
and Mark Zuckerberg from sending five cents a day to Save
The Children? Further, what form of humanitarianism does
not also rely on the assumption that, if atrocities
happen, it is because they are not known?[5] This critique
also avoids the questions: What networks are not banal?
And, to what extent does their power stem from their
banality, from the way they eviscerate details in favor
of interchangeable nodes and edges? Further, these
critiques miss what is arguably most intriguing about KONY
2012: its meteoric demise. It was a one-month wonder, kept
alive afterward by critical analysts, rather than adoring
fans. Its quick demise destabilized its creator's sanity;
Russell was captured on video naked on the streets of La

Jolla, CA. This video began with a warning, for it was deemed inappropriate for teenage girls, *KONY 2012*'s target audience. Overwhelmed by the glare of publicity generated by *KONY 2012*—by the constantly probing, hostile, yet adoring "we" it created—Russell overexposed himself in return. He gratuitously bared himself in order to show that he had nothing to hide.

KONY 2012's meteoric rise and fall reveals the ephemeral nature, as well as the dangers and pleasures, of contact and connection. *KONY 2012*, in the most banal way possible, reveals that we assume too much if we assume that we know what a connection is or does. It reveals the gap between network maps (which are always too early and too late) and our constantly broadcast gestures that are sometimes reciprocated and acknowledged, even as they are always imagined to be shared, that is, to connect.

=end KONY2012

1 Habitual Connections, or Network Maps: Belatedly Too Early

The imagination is now central to all forms of agency, is itself a social fact, and is the key component of the new global order.

—Arjun Appadurai[1]

Networks have emerged as a universal concept, resonating across disciplinary and political divisions, because they encapsulate neoliberal collectivity. By rendering the world into nodes and edges, networks both embody neoliberalism's vision of individuals as collectively dissolving society and foster analyses that integrate individual actions/tics into shareable trends/habits. Networks answer the dilemma posed by postmodernism—How to navigate an increasingly confused and confusing globalized world?—by diagramming allegedly unrepresentable interactions, from the spread of capital to affects. They capture individuals and (pre-)individual relations previously protected under the cover of the masses. They figure connections and flows—and constantly produce crises—by linking and breaching the personal and the collective, the political and the technological, the biological and the machinic, the theoretical and the empirical.

Although they enable individuals to cognitively map their relation to others, networks also confuse and obfuscate. Network subjects are both empowered and more precarious than ever. Neoliberalism despite—or more precisely due to—its rhetoric of empowerment has accentuated disparities in wealth, increased levels of individual debt, and depressed real incomes.[2] Neoliberal subjects—small s sovereigns—are always searching, rarely finding. Shifting from the zoom to the overview, from search term to search term, they defer and extend decisions; the end, like that mythic pot of gold, is never reached. At the same time, though, users' searches produce data that make users findable, even as they wander: searchability grounds findability and vice versa. Most perniciously, the rise of "Big Data"—linked to the recycling and integration of disparate databases across time and space—perverts the aim of cognitive mapping, for correlating seemingly unrelated individual actions has revealed larger connections, but this mapping has not enabled individual subjects to understand and change the system; rather, it has been used to preempt disruption and make users more predictable. Network maps render inert the dynamic systems they trace.

To begin to imagine networks differently, this chapter reexamines the grounding assumption of networks, namely that everything can be reduced to nodes and edges. In response to the basic question "What is a connection?," it demonstrates that connections are habits: habitual repetitions, which transform all other interactions—without which there could be no connections—into 'leaks,' that is, accidental (benign or malignant) contacts. It concludes by outlining how information is habit and the centrality of habit to data analytics and capture systems more generally.

Orienting Postmodern Disorientation

Networks end postmodernism. They counter pastiche with the zoom and the overview; they animate and locate 'wherever' architecture; they resolve multiculturalism through neighborhood predictors that bypass yet reinforce categories such as race, gender, sexuality; they replace postmodern relativism with data analytics. They do so by moving away from subjects and narratives toward actors and captured actions that they knit into a monstrously connected chimera. They imagine connections that transform the basis of the collective imaginary from 'we' to YOU: from community to an ever-still-resolvable grouping that erodes and sustains the distance between self and other.

Networks dissolve postmodern disorientation. In the late twentieth century, individual subjects seemed mired in a haze: 'placeless' architecture, frenetic commodification, unrelenting globalization, and media saturation. There was a growing consensus that it was now impossible for individuals to apprehend, let alone comprehend, their relation to the world around them, for, although the factors that determined their lives were global and inhuman, their means for navigating and negotiating their circumstances were painfully local and organic. Across the disciplines and across the political spectrum, the image of individual subjects caught in an overwhelming, unrepresentable, unimaginable, and chaotic global system was eerily repeated, along with calls for new tools and theories to map invisible constraints and consequences.

Cultural theorist Fredric Jameson made this argument most forcefully in his definitive diagnosis of postmodernism. Describing the navigational disaster that is the Los Angeles Bonaventure Hotel, he contended, "post-modern hyperspace … has finally succeeded in transcending the capacities of the individual human body to locate itself, to organize its immediate sur-roundings perceptually, and cognitively to map its position in a mappable external world."[3] Jameson linked this disorientation to nineteenth-century transnational capital, which undermined the relationship between indi-vidual perception and systemic truth (the truth behind nineteenth-century domestic tea rituals in England, for instance, lay in English colonial/trade relations with India).[4] Late capitalism exacerbated this gap between the authentic and the true, making it impossible for individuals "to map the great global multinational and decentered communicational network in which we find ourselves caught as individual subjects."[5]

Ulrich Beck, writing in Germany in 1986, diagnosed the emergence of what he called a "risk society" in similar terms.[6] According to Beck, we are moving from a system based on visible wealth (and thus class solidarity and humanly perceivable causality) to a self-reflexive modernity defined by invisible risks that produce "unknown and unintended consequences."[7] These risks, which can only be delimited scientifically, reverse the normal relationship between experience and judgment: rather than judgment being based on personal encounters and knowledge, it is based on a general knowledge that defies personal experience—that is, a "second-hand non-experience" that cannot be imagined.[8] Therefore, "a large group of the population faces devastation and destruction today, for which language and the powers of our imagination fail us, for which we lack any moral or medical category. We are concerned with the absolute and unlimited NOT, which threatens us here, the *un- in general*, unimaginable, unthink-able, un-, un-, un-."[9] Like Jameson, Beck concludes that this incapacity to imagine—that is, conceptualize or map—the threats around us prevents effective action.

In a less apocalyptic and thus less utopian manner, the sociologist Mark Granovetter, writing in 1973, also argued, "the personal experience of indi-viduals is closely bound up with larger-scale aspects of social structure, well beyond the purview or control of particular individuals."[10] To apprehend the relationship between personal experience and social structure, Granovetter produced one of the most influential social maps (see figure 1.1). This map tracks ties between individuals, where a tie represents an acquaintance. It transforms multiple interactions over time into lines that persist and represents individuals as static nodes. Through this figure, Granovetter countered the then dominant presumption that those with the most ties—that is, the social center—are the most powerful or influential.

The strength of weak ties

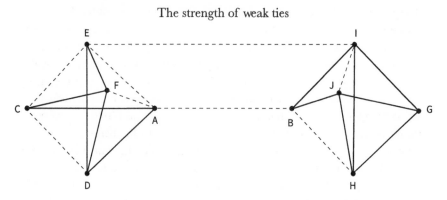

Figure 1.1
Mark Granovetter, Strength of Weak Ties (redrawn by Seungyeon Gabrielle Jung)

Rather, those with the most weak ties to others most effectively spread information and infection. Notably, this finding redefined what counts as powerful: disseminating rare information (information that cannot be readily gained elsewhere and thus can help one get a job, etc.) became more valuable than disseminating widely (i.e., spreading the same information to the greatest number of people). Granovetter's analysis moves us away from the mass, measured in terms of sheer force, to the N(YOU), gauged in terms of unique lines of influence and scarcity.

Jameson too posited new forms of mapping—of outlining and clarifying connections between locations and agents—as ways to reattach individual actions to global knowledge; but Jameson's vision was a vision. It was tentative and speculative. Drawing from geographer Kevin Lynch's argument that individual empowerment is linked to an individual's ability to cognitively map his/her surroundings[11] and from philosopher Louis Althusser's theorization of ideology as "the imaginary relationship of individuals to their real conditions of existence,"[12] Jameson posited cognitive mapping as a not yet imaginable form of political art, which corresponded to "an imperative to grow new organs, to expand our sensorium and our body to some new, yet unimaginable, perhaps ultimately impossible, dimensions."[13] Although cognitive mapping did not yet exist, Jameson viewed technology as having a special relation to it. Specifically, he suggested, "the technology of contemporary society ... seems to offer some privileged representational shorthand for grasping a network of power and control even more difficult for our minds and imaginations to grasp: the whole new decentered global network of the third stage of capital itself."[14] Our "faulty representations of some immense communicational and computer network" offer an outline, a figure, of how power (literally) flows among and through individuals.[15] Capital, for Jameson, was a network.

If for Jameson and Granovetter maps and networks once more connected the macro- and micro-level, the societal and the individual, for others—most influentially Gilles Deleuze and Félix Guattari—maps were central because they frustrated transcendental and totalizing discourses. Taking up the figure of the rhizome—a root structure characterized by connection, heterogeneity, and multiplicity—they posited the rhizome as a map that "fosters connections between fields, the removal of blockages on bodies without organs, the maximum opening of bodies without organs onto a plane of consistency."[16] A map did not trace anything; rather "the map is open and connectable in all of its dimensions. ... It can be torn, reversed, adapted to any kind of mounting, reworked by an individual, group or social formation. ... The map has to do with performance."[17] Drawing from Deleuze and Guattari, Bruno Latour has claimed that Actor Network Theory (ANT) frames actors "not as intermediaries but as mediators, they render the movement of the social visible to the reader."[18] **Thus, irrespective of political and intellectual differences, theorists have posited maps and networks—however defined—as key to empowering agents by making the invisible visible.**

This promise to capture seemingly invisible social and physical movements grounds the current predominance of networks as a theoretical tool. The maps Granovetter outlined have blossomed into dynamic representations used by corporations, researchers, and ordinary individuals to analyze almost everything, from friendship to contagious diseases. Affect theory, which often draws from the work of Deleuze to grapple with unconscious bodily reactions, uses the language of networks—of intensities, transductions, and connections—to trace affects that defy representation (affects lie both below and beyond individuals), yet enable communication.[19] The Internet is allegedly a rhizome. Interfaces and apps, from Google Maps to *Facebook.com*'s Graph Search to Twitter Analytics, offer us ways to trace the impact and spread of local connections. These acts of mapping—which allow users to track friends and followers and which offer users "the best routes" to familiar and unfamiliar destinations—are touted as empowering. Further, this logic of individual empowerment is embedded into the very premises of network analysis. As Mung Chiang asserts in *Networked Life: 20 Questions and Answers*, a textbook that serves as the basis for his popular *Coursera.com* Massive Open Online Course (MOOC) of the same name, networks operate most efficiently when nodes act selfishly. Describing Distributed Power Control (DPC), the algorithm that adjusts power among mobile phones in a given cell, Chiang comments that it reveals "a recurring theme in this book ... [that] individual's behaviors driven by self-interest can often aggregate into a fair and efficient state globally across all users, especially when there are proper feedback signals. In contrast, either a centralized control or purely random individual actions would have imposed significant downsides."[20] Here, Chiang describes feedback, which is central to modulation and optimization, as a supplement ("especially when"). He similarly frames interference as a "negative externality," as a factor that reveals that your happiness or success depends on another's action.[21] Hence, self-interested actions can be portrayed as central and determining only if other equally important relations are dismissed as secondary and/or accidental.

More critically, networks have been deployed across various fields to understand new power structures and new modes of individual and collective behavior in a society in which, as Margaret Thatcher infamously declared, "there is no such thing as society."[22] As Alexander Galloway and Tiziana Terranova have argued, control exists in and through seemingly unhierarchical network structures.[23] Galloway and Terranova posit a global "network culture," immanent to global capitalism, in which resistance is generated from within, either by hypertrophy or through the creation of common affects that traverse the network. Bruno Latour similarly contends that to do ANT, one must become an ANT: "a blind, myopic, workaholic, trail-sniffing, and collective traveler."[24]

These theoretical interventions have been key; but they have hardly solved the difficulties posed by globalization and late capitalism.[25] **We are now in a different and perhaps historically unique situation: we are forever mapping, forever performing—and so, we are told, forever empowered—and yet no more able to imagine, let alone decisively intervene in, the world around us.** Precarity, however liberating, is the dominant network condition, and mapping follows and amplifies networks. Maps may allow users to zoom out from the close-up to the overview—to see patterns and to move between scales rather than be mired in postmodern pastiche—but they seem to be always zooming and never changing, in part because users are, as in the *KONY 2012* video, simply zooming around a pastiche (see figures 1.2 and 1.3).

Further, the performance of mapping—even as it deterritorializes, opens new avenues and multiplicities, etc.—drives capitalism. As Ien Ang among others has argued, capitalism thrives on uncertainty and multiplicity.[26] Further yet, users have become so dependent on mapping technologies that they seem to be incapable of acting without these aids: the map has

Figure 1.2
Initial still of *KONY 2012* with the "invisible children"

Figure 1.3
Zoom out/pastiche of the "invisible children" in *KONY 2012*

eliminated the tour; representations of systemic truths now trump personal experience.[27] Tellingly, Google Maps and other mapping sites have become defaults for route planning, with links to these sites replacing personal directions on websites and invitations, despite the fact that, as Lisa Parks has pointed out, Google Maps stitches together old and often incorrect images (and thus, like figure 1.3, they are a realist compounding of post-modern pastiche). Finally, the emergence of networks has led to even greater financial insecurity: the financial meltdown in the early part of this century was due in part to the networked nature of "toxic assets."

To make sense of these new dilemmas, we need to ask: What exactly do we mean by "networks"? To what extent has the "solution" driven the prognosis? In what new ways have subjects been caught?

Networks: Projected and Existing

Networks are odd entities: they are both technical projections and naturally occurring phenomena. Modern networks stem from structures, such as electrical grids and highway systems, deliberately built to resemble nets (figure 1.4). However, networks are also empirically discovered phenomena. Systems biology, for instance, presumes the existence of networks in animals, from the genetic to the multicellular, which are discovered rather than simply modeled (figure 1.5). Similarly, ecology conceptualizes food webs and less lethal animal interactions—or, more precisely, the potentiality of these interactions—as networks (figure 1.6). These networks are portrayed as actually existing empirical entities, despite the fact that network analysis *replaces* real-world events with a reductive and abstract mathematical model.

As network scientist Duncan Watts admits, "rather than going out into the world and measuring it in great detail, we want to construct a mathematical model of a social network, *in place of the real thing*. ... The networks we will actually be dealing with can be represented in almost comic simplicity by dots on a piece of paper, with lines connecting them" (emphasis added).[28] This drastic simplification/substitution nevertheless enlightens because "although ... we inevitably miss features of the world that we ultimately care about, we can tap into a wealth of knowledge and techniques that will enable us to address a set of very general questions about networks that we might never have been able to answer had we gotten bogged down in all the messy details."[29] Networks are thus projections, theoretical diagrams (models based on past observations used to predict future interactions), empirical entities, and things to displace/analyze these real entities

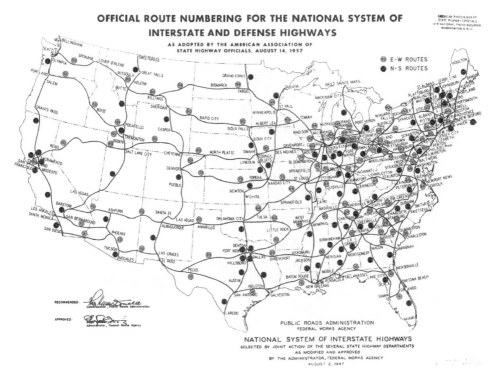

Figure 1.4
U.S. Highway Network, "Interstate Highway Plan, August 14, 1957," *Wikimedia Commons*, https://commons.wikimedia.org/wiki/File:Interstate_Highway_plan_August_14,_1957.jpg

and their messy details. Indeed, they compromise the distinction between the constructed and the natural, the theoretical and the empirical. Like Borges's infamous map, the map has become the territory.

Networks also spawn networks; they are as contagious as they are useful to diagnose contagion. Networks perforate the boundaries between the many disciplines that employ networks, from economics to media studies, from political science to biology. Every discipline, it seems, has discovered the network as a universal structure and thus found each other. The study of networks thus oddly mirrors its subject, making it even more difficult to separate network analyses from the networks they study. As Watts states, for the new science of networks to succeed, it "must become ... a manifestation of its own subject matter, a network of scientists collectively solving problems that cannot be solved by any single individual or even any single discipline."[30] It takes a network to resolve a network; networks generate networks, and networks succeed by spawning networks. (Perversely, the

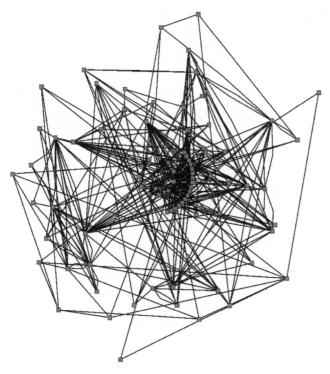

Figure 1.5
"A Thaliana Metabolic Network," *Wikimedia Commons*, March 2007, https://
commons.wikimedia.org/wiki/File:A_thaliana_metabolic_network.png   by   User:
TimVickers

most touted measure of "success" in 2014 for Teach for America, the non-
profit organization that puts recent college graduates mainly from elite
schools into "failing" public schools, was not the eradication of poverty
and inequality or the success of the students, but rather the network of successful alumni it had created.)[31] Networks are arguably as self-generating as
capital itself, hence their importance to mapping capital.

However dominant maps may be, 'liveness' defines networks. Networks,
drawn from communications systems, presume flows between nodes, so
that networks are 'alive.' Although networks are often technically called
"graphs,"[32] network theory differs from graph theory in its "view that
networks are not static, but evolve in time according to various
dynamical rules; and ... aims, ultimately at least, to understand networks
not just as topological objects, but also as the framework upon which distributed dynamical systems are built."[33] Network science, even as it relies
on "comic" simplifications,[34] tries to capture and understand events, such

Figure 1.6
Food Web Diagram, *Wikimedia Commons*, July 2012, https://commons.wikimedia
.org/wiki/File:Food_web_diagram.svg by User: LadyofHats

as catastrophic power outages and viral outbreaks, which fundamentally
redraw graphs. Critical uses of networks often try to separate the network
from network maps. Bruno Latour insists that a network is "an expression
of how much energy, movement and specificity our own reports are able to
capture ... it is a tool to help describe something, not what is described."[35]
Terranova insightfully describes networks—such as the Internet—not simply in terms of network infrastructure, but also in terms of information
flows, as do computer scientists such as Jon Kleinberg. Anna Munster in *The
Aesthesia of Networks* insists that a network is at least two things: an infrastructure map and a William Jamesian mosaic.[36] Emphasizing the latter,
Munster argues that the pulsing of energy and affect—the network
experience—cannot be reduced to nodes and edges, for networks are *about*
edging: pulsations that frustrate neat separations and create sticky connections between the molecular and the molar.

   Apprehending the difference between the experience of flows and maps
is important, and understanding contact as generating nodes is central to
reimagining networks. However, the double-faced nature of networks, as

both trace and flow, is exactly what makes networks so compelling. Networks spatialize the temporal by tracing and projecting: by being both too early and too late. It is not an issue of choosing one over the other, but rather, like Donna Haraway's cyborg, a call to see both at once.[37]

Networks are both science fiction and the historical present. They describe future projections as though they really existed; they relay past events as if they were unfolding in the present. The UNIX command traceroute, which allegedly traces our data packets, illustrates this tense nicely. The traceroute tool sends out a series of packets with increasing TTL (Time to Live) values, starting with one hop. Whenever the packet 'dies,' the router at which the packet expires sends a message back to the originating machine; but, since packets can take different routes through the network and since many routers will refuse TTL error messages, this 'trace' is not entirely reliable. Through timed TTL settings, traceroute offers us a pastiche of packets to map what allegedly has been, is, and will be.

Networks also render time into space—they spatialize the temporal—by portraying polyvalent interactions as direct lines of contact. CDMA (Code Division Multiple Access) exemplifies this logic. At first, the idea that cell phones comprise a network seems strange, since all transmitters and receivers send their signals into the air. Technically speaking, a cell is not a single phone, but rather the area covered by one cell phone tower. To create a network, cell phone protocols such as CDMA use different frequency bands and individuating codes to create identifiable and legitimate connections: to encode and decode signals between a single transmitter and the receiving tower. These algorithms also classify interference, generated by the very towers and devices that transmit and receive, as well as by the air through which signals travel, as external. Through this, diffuse clouds of interacting signals become neat network maps in which nodes connect directly to other nodes.

These technical examples are important, not because technology underpins or determines everything else, but rather because they can help explain the attraction of the odd double-edged (or edging) power of networks. To return to Jameson's insight that contemporary technologies offer a "privileged representational shorthand for grasping a network of power and control even more difficult for our minds and imaginations to grasp: the whole new decentered global network of the third stage of capital itself,"[38] they do so because they reveal how networks are imagined and created. Technical networks cut continuous space and time into slices of connectivity: they project links and cut noise to create neat lines between transmitters and receivers.

Link: Leak

Imagining networks in terms of nodes and clean edges creates traceable paths, but it also raises the specter of noise and leaks, that is, of everything that distorts these connections and of all that "mistakenly" connects. It classifies as malevolent or accidental interactions that do not simply compromise connections, but also make them possible. For example, Claude Shannon's foundational definition of information as entropy classifies noise as a separate source, added to the information channel (figure 1.7).[39] Electromagnetic noise, however, does not always come from the outside, for it is also generated by the very act of transmission: there is no separate origin for this uncertainty. In terms of networks, leaks are not accidental; they are central. Without leaking information, there could be no initial connection.

At a very basic level, our networks work by 'leaking.' A wireless network card reads in all the packets in its range and then deletes those not directly addressed to it. These acts of reading and erasure are hidden from the user, unless she executes a UNIX tcpdump command or uses a packet sniffer in promiscuous mode so that her network card writes forward these packets to the computer's central processing unit (CPU). (Remarkably, packet sniffers— network analysis tools developed to help sysops diagnose networks that are always failing—are now illegal to install on computers you do not own in several countries). The technical term "promiscuous mode," however, is a misnomer: whether or not you make your network card promiscuous, it acts promiscuously. A network card only appears faithful to its user because

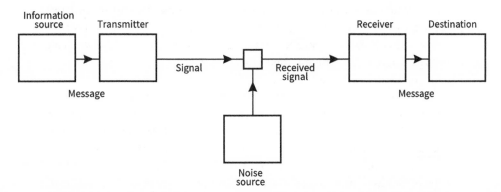

Figure 1.7
Shannon and Weaver's Diagram of Communication (redrawn by Seungyeon Gabrielle Jung)

it discreetly erases—that is, does not write forward—its indiscretions. (Tellingly, "monogamous mode" is not a technical term; a truly monogamous network card would be inoperable.) To return to the CDMA example described earlier, within any given cell everyone's mobile phone receives the same signal, and other people's mobile phone signals interfere with any one person's signal; one user's signal is another's noise. **Our devices, our computers, constantly leak. They are wonderfully creepy.**

All this is hidden by the neat tracings of traceroute and other network mappings, which erase the existence of crosstalk and other wonderful, generative effects of electromagnetic interference. This leaking is not accidental; it is central. Without this constant exchange of information, there would be no communication, no Internet. Crucially, though, this leaking does not automatically make the Internet a great surveillance device. As I argue throughout this book, it takes a lot of work—the National Security Agency, corporations that encourage "real names" and other unique identifiers, massive server farms, etc.—to render the Internet into PRISM (Planning Tool for Resource Integration, Synchronization and Management). The rapid reading, writing, and erasing that drives TCP/IP must be supplemented by a technics and politics that seeks to store everything, to make memory storage.

Most perversely, given the ephemerality of electromagnetic media, computers have erased the difference between memory and storage: users now store things in memory, rather than storing memory traces.[40] This storage, however, is not outside this leaking. Even seemingly sealed, not connected, our computers constantly leak: they write to read, read to write, erase to keep going. Without this leaking, nothing would remain, at least not digitally, because **now in order to remain, nothing remains, and so nothing remains even as everything does.** As I explain in the next chapter, to 'store' something digitally, one often destroys what actually exists and can persist far longer than digital media, such as paper and film; when 'saving' a file, one writes over an existing one.[41]

To engage with the wonderful creepiness of new media, we need to rethink and reenvision connection. Imagined connections are habitual actions. Habits, as Deleuze writes in *Difference and Repetition*, make series and seriality—that is, difference—conceivable as generic relations.[42] To return to the analogy of the food web as a network, what matters is not that a fox in Yellowstone with tag 104 ate a rabbit (there are too many rabbits to tag), but rather that foxes habitually eat rabbits. Each line is a potential interaction based on repeated past interactions. This "comic" connection relies and thrives on repetition, past and future. In order to be transmitted, signals must be repeated. Signals that are not repeated or repeatable 'die.'

Networks, for all their preoccupation with singular or virtual events that fundamentally alter network maps, rely on continual repetitions—or the possibilities of repetition—which they portray as static ties. Repetition makes it possible to imagine and elucidate connection, as well as link humans and machines together as beings that repeat.

Connecting Habits

Imagined connections are habits: they are projected links based on frequent and potential repetition. According to the philosopher David Hume, who first rigorously formulated the relationship between causality and probability, habit establishes causality; it grounds the "always."[43] Imagined connections and edges—things that remain—are traces of habits. In terms of social networking sites, the strength of a friendship—its weight—is gauged by the frequency of certain actions. More strongly: **information is habit.** Habit resonates with two seemingly unrelated meanings of information: one, the archaic definition of "information" as the formation—the training—of individuals (to inform was to form); and two, Shannon and Weaver's definition of information as communication that lies beneath meaning.[44]

It is important to remember that, although networks are imaged as graphs (see figure 1.8), they are analyzed numerically as matrices. The networks in figure 1.8 can be represented mathematically as the adjacency matrices A, in which:

$A_{ij} = \{1$ if there vertices i and j are connected; 0 if they are not$\}$

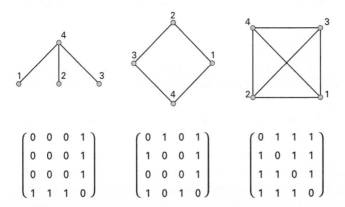

Figure 1.8
Matrix representations of networks: Eric W. Weisstein, "Adjacency Matrix," *Wolfram MathWorld*, http://mathworld.wolfram.com/AdjacencyMatrix.html (redrawn by Seungyeon Gabrielle Jung)

In many matrices, rather than a simple 1 for a connection or 0 for not, the edges are weighted in terms of significance—for instance, in relation to the number of groups nodes have in common; the weights thus reveal the likelihood of being connected in the future. Intriguingly, two of the most popular network algorithms are based on academic measurements (themselves a result of the bureaucratization of the university and the quantification of research): the citation index (which lies at the heart of *Google.com*'s search algorithm) and the coauthor index (which has inspired various affiliation network analyses). In a citation network, the more citations a paper has, the more important it is considered to be; a coauthor network is a bipartite network that tracks both papers and authors to determine the level of affiliation between both entities (that is, to see how 'alike' various nodes are).[45] Collaborative filtering algorithms such as the Netflix prize algorithm, discussed in chapter 3, move from analysis to prediction by building 'neighborhoods' based on strong similarities (and differences) between the reactions of users to films.

Mathematically speaking, habit and information are linked via the notion of probability (habit and information are also linked more philosophically, as I argue in chapter 4, as forms of repetition beyond and beneath meaning). Shannon defined information as the degree of freedom within a message.[46] For example, if there are eight possible messages that can be sent, the information of that set of messages would be four bits (eight in binary is 1000). Information is crucial to determining the capacity of a channel, since a channel's capacity must equal the information of the set of messages to be sent down that channel, at the very least. Usually, not all messages are equally probable, and so information is the set of weighted probabilities:

$$H = -\Sigma p_i \log p_i.$$

In this equation, H is the measure of information, which is the negative value of the sum of the probability of any given symbol (p_i) multiplied by the logarithm of p_i.

Habit is key to determining probabilities, for habits render past contingent repetitions into anticipatable connections. Given David Hume's groundbreaking work on anticipation and repetition, it is no surprise that Hume is the favored philosopher of Big Data analytics; references to him appear in popular venues such as *Wired* magazine and in PowerPoint presentations made by those advising the U.S. Intelligence Community.[47] Although these references rarely engage Hume's work seriously, they highlight the importance of habits to understanding how causality, correlation, and anticipation work in the era of the inexperienceable experience. Deleuze, reading Hume's *Experience and Subjectivity* (a text that would have a profound effect on Deleuze's later work), explains how Hume's theory of causality links experience to habit as follows: causality, Deleuze writes, does not proceed "on the basis of certainty" (it is not based on "intuition or demonstration"), but rather on the basis of "probabilities."[48] This does not mean that causality is derived from probability, but rather that causality forms gradually and is the result of habit, which presupposes, even if it does not coincide with, experience. According to Hume, "experience is a principle, which instructs me in the several conjunctions of objects for the past. Habit is *another principle*, which determines me to expect the same for the future."[49] Experience presents cases of constant conjunction to the inspecting mind, but "repetition by itself does not constitute progression."[50] Habit is the root of reason. Habit allows the mind to transcend experience, to reason about experience "as it transforms belief into a possible act of the understanding."[51] Causality is thus both "the union of similar objects and also a mental inference from one object to another."[52]

Crucially, though, habit and experience are not—and do not have to be—unified. Habit, for Hume, can falsify experience: it "can feign or invoke a false experience, and bring about belief through 'a repetition' which 'is not deriv'd from experience.'"[53] These beliefs however inevitable are, Hume stresses and Deleuze underscores, illegitimate: they "form the set of general, extensive, and excessive rules that ... [are called] *nonphilosophical probability*."[54] To correct these beliefs, Hume explains, the understanding intervenes through a corrective principle that restrains belief to the limits of past experience—to the "rules of *philosophical probability* or the calculus of probabilities" so, although "the characteristic of belief, inference, and reasoning is to transcend experience and to transfer the past to the future; ... it is still necessary that the object of belief be determined in accordance with a past experience."[55] According to Hume: "[W]hen we transfer the past to the future, the known to the unknown, every past experiment has the same weight, and ... 'tis only a superior number of them which can throw the balance on any side."[56] This passage clarifies the challenges posed by 'unimaginable' risks, for **in a risk/postmodern society, understanding can neither draw from experience (the conjunctions of objects from the past) nor be corrected by it.** This, however, does not mean that experience and habit are irrelevant, because, in a networked society, the link between habit and correction itself becomes of the second order. What matters now—what cognitive mapping has become—is divining causal relations not between things that repeat successively but rather across time and space. Philosophical probability, in other words, expands beyond an individual's experience to draw from experiences of people 'like you'; through data analytics, your history becomes YOUR history.

Big Data as Cognitive Mapping

Big Data, through its integration of second-order habit, offers a form of cognitive mapping that allegedly sees all, by ignoring causes. Most boldly and controversially, *Wired* editor Chris Anderson has asserted, "the data deluge makes the scientific method obsolete."[57] Big Data ends the need for hypotheses and for theory; humanists are not the only ones discussing "the end of theory."[58] As many have noted, Anderson's article has many problems: it places statistical analysis outside theory, as though statistical analyses, designed to recognize significant patterns, did not themselves draw from theory; it assumes that data can speak for itself (as Lisa Gitelman has shown in her edited volume *"Raw Data" Is an Oxymoron*).[59] Regardless, Big Data, or data analytics more generally, challenges our common perceptions of causality because clearly noncausal relations—that is, seemingly accidental or nonchronological relations—seem to be better predictors of future behavior than so-called causal ones.

In their popular book *Big Data*, Viktor Mayer-Schönberger, professor at the Oxford Internet Institute, and Kenneth Cukier, data editor at the *Economist*, have argued that Big Data fundamentally challenges the efficacy of causality because it shows that correlation trumps causality: the 'what' matters more than the 'why.'[60] Mayer-Schönberger and Cukier offer now widely canonized cases as evidence: FICO's "Medication Adherence Score," which determines how likely patients are to take medications regularly based on information such as car insurance payments; and Target's "pregnancy prediction" score based on the purchase of vitamin supplements and unscented lotions.[61] The Target example has become the poster child for Big Data, for Target allegedly predicted the pregnancy of a teenage girl before her father knew of it based on her purchases. After accosting a Target manager for encouraging his daughter to become pregnant by sending her baby-related coupons, the father of the girl later apologized when he realized that his daughter was indeed pregnant. As Charles Duhigg further explains, the pregnancy predictor is an especially important indicator for Target, because pregnancy is one of the few transition periods in which shopping habits can radically change.[62] Based on these cases, Mayer-Schönberger and Cukier contend that we need to give up on causality not only because knowing what (e.g., a person is pregnant) is more important than why, but also because knowing why is often "little more than a cognitive shortcut that gives the illusion of insight but in reality leaves us in the dark about the world around us."[63] They further argue that we need to let go of our penchant for accuracy. In terms of Big Data, accuracy is not needed, for it is better to have a lot of noisy data than a smaller set of accurate data. Despite chaos theory, knowing about that butterfly's flutter in South America does not matter.

An analysis of habit enables a more nuanced view of the changing relation between correlation and causality. On at least three levels, data analytics are about habits: one, they focus on habitual actions, such as buying lotions and vitamins; two, based on this analysis, they seek to change habits, especially by focusing on moments of 'crisis'—moments of state change—such as pregnancy; and three, they 'replace' causality with correlations **between** habits. That is, correlations between correlations rather than correlations between repeated series of events are key. If, for Hume, anticipation stemmed from habit—where habit is a belief based (or not) on experiential repetition—for Mayer-Schönberger and Cukier, anticipation stems from correlating habits such as paying car insurance and taking medication regularly. This second-order correction, which depends not on individual past experience but on collective actions, presupposes a massive amount of data, seemingly impossible to analyze analytically, that also makes sampling a thing of the past. It replaces errors drawn from sampling with errors in data, which are allegedly miniscule compared to the mass of data available. Through this, singular actions become indications of collective patterns rather than exceptions. That is, in the world view of data analytics, there are no singular actions, and thus many false positives.

Big Data and social media are also linked to habit at a fourth level. The Internet as Big Data, as I elaborate in chapter 3, is not a natural or inevitable outcome; if it is possible now to have such data sets, it is due to a politics and a practice of memory as storage, which creates fairly robust long-term data trails. This politics and practice rely on users becoming habituated to owning their connections so that a relatively solid longitudinal data set, which follows individuals and individual actions through time, can emerge. This process is remarkable, for when the Internet was first conceived, IP addresses, even when fixed, were not viewed as permanently tethered to a computer, let alone a user. With the advent of changes to IP addressing, and more importantly the emergence of cross-platform logins, cookies, and other means of tracking through 'unique identifiers,' it has become easier to tie users to their actions. This traceability has entailed the massive rehabituation of individuals into authenticated users through the expansion and contraction of privacy via notions such as 'friends' on social media; that is, through modes of leaking that both undermine and buttress walls that supposedly protect and secure.

Not accidentally, the correlations exposed and exploited by many of the consumer uses of Big Data focus on the amplification of consumer behavior: if you have bought this, you probably also want to buy that. The goal is to program customers to act in certain ways (or to predict present conditions or future habits), based on habits already contracted. Corporate uses of Big Data are not interested in changing behavior radically (or in prevention), but rather in amplifying certain already existing behaviors and preempting others. It is not interested in causes, but rather proxies (such as unscented lotions and vitamin B for pregnancy). Also, the fact that it takes data analytics to realize that human female procreators suffer from morning sickness is dumbfounding and raises questions about corporate hiring. How much less expensively could Target have figured this out, had it had more women in leadership positions? Further, as the medical insurance examples reveal, Big Data can lead to rational yet unjust conclusions: the tie between car insurance and regular adherence to a regular medication regimen can further exacerbate inequalities by making the urban poor pay more for their health insurance. In other words, by finding seemingly unrelated correlations, Big Data can aggravate existing inequalities and lead to racist and discriminatory practices, justified through the use of seemingly innocuous proxies. Through these proxies, the allegedly "coarse" and "outdated" categories of race, class, sexuality, and gender are accounted for in unaccounted ways.

As Oscar Gandy, Jr. has asserted, these systems are not about fairness or justice, a problem that becomes more pressing in relation to algorithms based on correlations, which intelligence-gathering services use to create lists of probable suspects and potential terrorists.[64] As Gandy notes, there are several problems with these systems: from the impact of invalid or inaccurate data to the difficulty of discerning which correlations really matter (more on this later) to the untrackability of false positives. In a system designed to preempt action, it is impossible to know how many innocent people have been falsely detained or arrested. Further, this type of system seeks quiescence by quickly figuring out who to target and who to leave alone, so that the smallest number of people are subjected to invasive security measures and discipline. It is "transparent," where transparency, as Antoinette Rouvroy has argued, refers not to user knowledge of the system, but rather to user ignorance.[65] Most insidiously, as Rouvroy points out, this type of system seeks to eradicate preemptively everything in the human being that remains uncertain or virtual. In addition, by relying on users' actions rather than their words, it denies subjects the ability to give an account of themselves—actions are always presumed to trump words; the body never lies.[66] Rouvroy also contends that by recording everything and relying on nonhuman forms of perception, such systems deny humans the ability to forget and thus to create new norms. Crucially, these systems create a false sense of security: as the 2013 bomb blasts in Boston revealed, relying on statistical patterns and probabilities leaves one vulnerable to more improbable suspects, to whom they grant more safety. The light of Big Data creates big shadows through its very mechanism of capture, which shapes the reality it allegedly mirrors by depending on past data to 'pass on' data.

The Logic of Capture

Imagined connections depend on the logic of capture. Capture systems, which undergird data analytics and optimization, buttress the relation between leak and habit. The logic of capture lies at the heart of the oddly performative nature of networks.

Capture systems, as the computer scientist Philip Agre has explained, drive tracking systems, such as active badges and barcodes, that allow an activity (like shopping, lecturing, or driving a car) to be broken down into discrete units, which can then be articulated (strung, spoken) into various grammars and schemes for optimization and normalization.[67] According to Agre, capture systems are sociotechnical systems that have deep roots in the practical application of computer systems, for "capture" denotes two important functions in computer science.[68] Crucially, although Agre does not remark upon this, these different notions of capture also represent two ways of 'passing on' data. The first definition of "capture," the most common, "refers to a computer system's (figurative) act of acquiring certain data as input, whether from a human operator or from an electronic or electromechanical device."[69] For example, data is captured when an item is scanned at a cash register. Importantly, this data is usually not acted on immediately but rather 'passed on' to a database to be used later: a capture system saves traces of actions to create histories of past and future actions. The other use of "capture," common among AI (artificial intelligence) researchers, "refers to a representation scheme's ability to fully, accurately, or 'cleanly' express particular semantic notions or distinctions, without reference to the actual taking in of data."[70] An AI program has successfully 'captured' a behavior when it can mimic an action—like a typical retail transaction—without having to sample the actual movement. It works, in other words, when it can 'pass on' data because data is not necessary. So, there is an odd paradox at the heart of capture systems: they are both representation and ontology, data and essence. A capture system works when it no longer needs to capture, or rather when it can capture in the second sense of AI systems, without necessarily capturing data in the first sense of registering a trace of some action. Perversely, the dream of capture is the eradication of capture, an eradication that can only be confirmed as accurate by continued capture.

Agre maintains that the metaphor of human activity as a kind of language is deeply ingrained in these systems, for they impose on users a grammar of action articulated through a project of empirical and ontological enquiry.[71] This capture, which is framed as a discovery, Agre emphasizes, is an intervention, for it seeks to rewrite the procedure it 'discovers.'[72] Because these systems are focused on action and optimization, Agre also aligns capture systems with marketization and democratization. Drawing from Ronald Coase's theory of the firm, Agre states that once a grammar of action has been imposed on an activity, discrete units and individual episodes of activity are more readily identified, verified, counted, measured, compared, represented, rearranged, contracted for, and evaluated in terms of economic efficiency. Capture systems actively restructure what they allegedly discover, that is, they allow not only for corporate imposition but also for a certain empowerment, based on the marketization of all transactions. Capture systems buttress the economization of all interactions central to neoliberalism.[73]

Agre argues that capture systems, because they are focused on grammars of action, are not surveillance systems. He explains that surveillance systems employ visual/territorial metaphors and are centralized systems linked to state power; capture-based systems, in contrast, are linked to private corporations. Capture and surveillance systems also differ in terms of their base unit of analysis. In a capture system, the base unit is an action, or change of state, rather than an entire person. A capture system enables a finer grid by presuming and enabling mobility, for, in order for something to be captured, it must be in motion. There must be a change of state; things must be updated in order to register. Because of this constant activity, people who engage in heavily captured activity have a certain freedom, namely, free creation within a system of rules.[74] They can optimize their actions, so that their effort is decreased or their recorded productivity is increased; they can become more rather than less skillful. In more cynical colloquial terms, users can game the system. They can, based on what and how things are captured, work to score highly in those categories. For instance, in 2015 buses in a certain city in the United States were tracked and bus drivers assessed according to their ability to stick to a prescribed schedule. This ability, however, varies radically not only with traffic conditions but also with the number of passengers: the later the bus is, the more passengers there are waiting to be processed, and so the bus becomes later and later. In order to compensate, bus drivers would often not charge passengers at busy stops. This empowerment, of course, fosters a game of cat and mouse: buses in that city were then equipped with surveillance cameras in response to this "freedom" of the drivers. More darkly, the costs of empowerment within capture systems are clarified by the deleterious effects of standardized testing on education within the United States. Under the rubric of "no child left behind," schools, which are ranked according to their students' scores on these tests, have become increasingly focused on teaching students how to pass standardized tests over all other skills. This empowerment—via grammars of action—also silences by denying the importance of language. If language is central to the act of claiming rights, the rejection of spoken language in favor of parsed actions, as Rouvroy points out, has troubling consequences for the future of democracy.[75]

Although Agre separates capture and surveillance systems, it has become clear not only that this division is false, but also that the separation of state from corporate power, which grounds this division and certain naïve conceptions of neoliberalism, is fictional at best. Edward Snowden's revelation about the extent of cooperation between the U.S. National Security Agency and corporations such as Google and Facebook reveals this nicely. Again, the protests by these corporations against the U.S. government ring false, not only because of the history of their cooperation with the state, but more importantly because the value of the NSA's data stems from corporations' insisting on real names and unique markers. Surveillance has become a co-production.

Rather than focus on these dramatic revelations, which should have surprised no one, the rest of this book addresses this muddied relationship between public and private in terms of habits. Capture systems are all about habitual actions. They seek to create new, more optimal habits; they record habitual actions in order to change them. Surveillance, and the older notions of privacy, remain through our habits, which are also central to discipline. They remain in how we situate, or inhabit, our actions. Contra Agre, surveillance and capture also intersect as processes of intervention. The next chapter examines the confluence of surveillance and capture through the temporality of networks as crisis, crisis, crisis. It reveals the centrality of crisis to the evolution of capture systems—to the rendering of events into mappable networks and programs.

To conclude, I want to emphasize—with and against Agre—that capture systems and their logic of tracking offer possibilities for user actions that do not necessarily disempower, because they also offer a means of imagining and inhabiting the capture. The term "glocal" may be insipid, but it does reveal a desire for a traceability and responsibility. Some of the products and ways of achieving this seem dubious: from applications that reveal how much 'slave labor' was involved in producing an item to attempts to raise chickens in Queens, New York, so that children can know from where their eggs come.[76] These efforts are suspect not only because of the assumptions they make (for example, what counts as 'slave labor'), but also because their scope seems painfully limited to individual actions. Indeed, capture and habits, by reifying individual actions and empowerment, ignore the question of larger structural changes and impacts. Although true, the desire for traceability also creates desires for contact that, as I argue in chapters 3 and 4, can lead elsewhere. To read 'voice' in systems designed to deny voice, we need to 'listen' for actions that are embedded in infrastructures and settings and learn to speak through these actions. The desire to capture and to draw connections, maps, and models can open a future that defies what is captured, but only if we dwell in the disconnect between map and action, model and future—only if we occupy the collective chimera we are offered and become characters, not marionettes, in the ongoing drama inadequately called Big Data.

Habit + Crisis = Update

=TXTMob

In late August 2004, New York City hosted the Republican National Convention (RNC), in which then-President George W. Bush was formally accepted as the Republican candidate for that year's presidential election. Given the catchy theme, "Fulfilling America's Promise by Building a Safer World and a More Hopeful America," it was, like all recent U.S. national conventions, a pro-forma, made-for-TV event that sought to use free television news time to generate momentum for the campaign. The choice of New York City was symbolic, meant to remind U.S. citizens of the 9/11 attacks and the ensuing unity. By holding the convention in one of the most liberal cities in the United States (one that, besides electing Republican mayors, almost always elects Democratic representatives), it also asserted the reach of the Republican Party.

This reach, however, was disputed by the constant protests and marches that coincided with the convention. The main march on Sunday, August 29, organized by United for Peace and Justice, drew approximately 500,000 people.[1] While these protests do not compare with the traumatic and violent protests outside the 1968 Chicago Democratic National Convention (DNC), they nonetheless hold the record for the most arrests during a political convention: over 1,800.[2]

Although less than a week in duration, the protests (and the police reaction) had been planned at least nine

months in advance. This advance preparation was due not
only to the level of pent-up frustration over
the war in Iraq but also to the police adoption of the
so-called Miami model.[3] Named after the law enforcement
reaction to protesters at the 2003 Free Trade Area of the
Americas summit, the Miami model is "characterized by
restricting public access to large parts of the city,
preemptive arrests of activist 'leaders,' widespread use
of nonlethal weapons including tear gas, pepper spray and
rubber bullets, and the use of mass arrests or 'sweeps'
that often includes the detention of law-abiding citizens
who are released without charge."[4] The Miami model has
been condemned as de facto martial law and for
criminalizing political protest.[5] In reaction to this
model, protesters have adopted a tactic of 'swarming': of
decentralized and fluid actions that quickly appear and
then disappear.

The 'swarm' used TXTmob, a tool developed by the Institute
for Applied Autonomy (IAA) in collaboration with the
protesters, to mobilize and organize.[6] Tested initially at
the Boston DNC (which tried to sequester protesters within
a "Free Speech Zone," enclosed by chain link fencing and
razor wire), this application enabled protesters to send
one-to-many text messages.[7] It was, in essence, a listserv
for text messages. This was not the first time wireless
technology had been used to organize protests: text
messaging was deployed more spontaneously in protests such
as the 2001 People Power II demonstrations in Manila;
two-way radios had been used previously by U.S.
protesters. TXTmob, however, offered a more formal many-
to-many messaging system that, unlike two-way radios, had
a wide message range and user base.

TXTmob enabled four types of groups: (1) public and
moderated, (2) public and unmoderated, (3) private and
moderated, (4) private and unmoderated. The first was used
to create and coordinate flash mobs, as well as to
disseminate "actionable information" (such as "Police
moving fast Westbound on 23 St toward the bike bloc")from

trusted sources.[8] Key to TXTmob's operation was the
forwarding of credible messages (perhaps paradoxically,
messages were often deemed credible based on who was
forwarding them). Unmoderated public groups were used to
create open forums; they were also unexpectedly employed
by remote users to participate 'virtually' through
messages of support and real-time reporting of mainstream
media coverage of the protests. The private groups were
utilized by medics in a dispatch model and also used for
more personal meet-ups and information sharing.[9]

According to Tad Hirsch and John Henry, developers of the
software, TXTmob offered a mode of 'topsight' or
'cognitive mapping': it provided a "'big-picture' view of
what's going on," allowing activists and journalists
to create a "mental picture" of all the actions taking
place in the city (not surprisingly, it was also used by
nonprotesters to navigate street closures).[10] Hirsch and
Henry also underscore the fact that it enabled "new forms
of participation and collaboration among activists,"
including more spontaneous actions, such as "kiss-ins,"
and more remote ones, such as off-site participation.[11]
Arguably, the most important cognitive mapping happened
after the event: Hirsch, who was working on a PhD
dissertation at the time, saved data from TXTmob (users,
though, were given the option of deleting their messages).
He was subpoenaed in 2008 for the TXTmob records.[12] As
Hirsch notes, this subpoena seemed oddly redundant, for
protesters had assumed police infiltration of TXTmob, and
of the protests more generally. Perhaps the possibility
of various TXTmob privacy settings fostered modes of
activity that belied this knowledge (again, ideology as
habit not knowledge).

The efficacy of TXTmob and the RNC protests more generally
remains an open question. Many derided the swarmlike
protests as "carnival[esque]" and demeaning to the
tradition of social protest.[13] Conservatives encouraged
delegates to take pictures of the "freak[s]" and to post
them online in order to exemplify the difference between

these "communist" activists and mainstream Americans.[14]
Remote participation was similarly dismissed as
"voyeuristic" and "risk-free." More concretely, despite
the protests, the Bush-Cheney team was reelected in 2004.
Even more concretely, the protests escalated the Miami
model: again, more protesters were arrested during this
convention than during any previous political convention
in the United States. (Almost ten years afterward, the
city of New York would agree to pay $18 million to settle
the civil rights claims of many of those arrested.)[15]
Given the police participation on TXTmob, those who used
TXTmob were also probably subject to surveillance long
after the protests. TXTmob thus increased the longitudinal
and geographical range of police investigations. TXTmob
itself seems to have disappeared through its success: it
is allegedly one of the inspirations for *Twitter.com*.[16]

TXTmob, however, is important because of, rather than
despite, these contradictions. These failures and
successes expose the dangers in using what would become
'social' media on 'personal' devices to organize
networked, dynamic political protests. The contaminations
of the serious and the innocuous, the performative and the
oppositional, the here and the there, and the real time
and the stored are the point. By seeking to politicize
flash mobs, TXTmob embraced the challenges posed by the
spread of privatization and the contraction of the private
that 'personal' devices encapsulate and foster.[17] Most
importantly, TXTmob revealed the possibilities and
limitations of using new media in real time as a way to
control and create crises. TXTmob literally created and
directed "turns" by giving directions to protesters to
turn right and left, to run or hide; and directions to
police officers about where to find them. TXTmob also
sought to overturn: to overturn the banal spectacle of the
RNC with its own festive one and to overturn the Bush-
Cheney doctrine of preemption through preemptive, just-in-
time actions.

TXTmob lives on in all the political crises tracked and
managed via *Twitter.com*, as well as within the increasing

transformation of catastrophes into crises.[18] With the adoption of a seemingly innocuous technology based on "idle talk"—*Twitter.com*—**there are no more catastrophes, only crises.** Natural disasters and technological failures are not things that simply happen to people; they are things that demand decision, action, and a constant stream of updating information, for public participation is the new norm in crisis management. Social media usage increases during moments of crisis.[19] Governments and corporations must actively update the public if they want to avoid appearing as withholding or guilty, since innocence is no longer presumed. At best, this creates a sense of being "in touch." At worst, it foments an acceptance and desire for surveillance. As Anil Dash (appropriately) tweeted in response to the shocked and incredulous public reaction to the missing Malaysian Airline Flight 370, "a triumph of surveillance culture that people seem downright disappointed a plane can disappear."[20] After Edward Snowden's revelations, **the fact that a government does not know everything is unacceptable, even to those to whom a government knowing everything is unacceptable.** Ignorance or silence indicates conspiracy and the beginning of an endless cat-and-mouse game of explanations and evasions in which authority is challenged and reestablished.

TXTmob and its legacies and predecessors also reveal the extent to which crises, political and otherwise, disrupt habitual modes of action and thus enable the wide-scale adaptation of novel technologies and the emergence of alternative voices. Governments and corporations must respond immediately because crises—the sometimes intense, sometimes banal collision of affect and consciousness—allow new authorities to emerge.[21] Actors take advantage of crises in order to attack chronic conditions and create new defaults.

TXTmob's most instructive legacy, however, is its apparent failure as a legacy. Unlike *Twitter.com*, there were no political revolutions and movements named after it; unlike

the protests in Moldova, the RNC protests were not named
the "TXTmob conventions." This is not simply because
TXTmob was not widely known outside the groups of
protesters and police; after all, Moldova hardly had a
large group of national Twitter users at the time of its
protests. The failure of TXTmob as a legacy reveals the
importance of imagined connections. Moldova and Tunisia
became Twitter revolutions because U.S. citizens could
imagine their technologies in action: *Twitter.com* was
fascinating because it created links between me and YOU,
which infiltrated every corner of the world, allegedly
spreading democracy as it did so. TXTmob serves as an
intriguing negative model, which exposes the conditions
necessary for the complete eradication of the political
in favor of the technological, that is, for the erasure
of the local issues in Moldova—the erasure of the
political problems and location—in an allegedly global yet
trackable technology.

=end TXTmob

2 Crisis, Crisis, Crisis, or The Temporality of Networks

Only a crisis—actual or perceived—produces real change. When that crisis occurs, the actions that are taken depend on the ideas that are lying around. That, I believe, is our basic function: to develop alternatives to existing policies, to keep them alive and available until the politically impossible becomes politically inevitable.

—Milton Friedman[1]

[Cruel optimism]: what happens to fantasies of the good life when the ordinary becomes a landfill for overwhelming and impending crises of life-building and expectation whose sheer volume so threatens what it has meant to "have a life" that adjustment seems like an accomplishment.

—Lauren Berlant[2]

Networks are made out of time: the chronic time of habits (memory) and the punctuating time of crisis. Unfolding in real time, habitual repetition grounds ties; crises break and create new ones. Crises—turning points, events that demand decisions—ensure that networks differ from graphs; it makes them alive and volatile. Crises undo habituation and undermine autonomy: they turn habits into addictions. Just when we are finally accustomed to something, it changes. Habit + Crisis = Update; program (X) + exception (X) = program (X + 1).

Crises drive the endless optimization of capture systems; because the capture systems discussed in chapter 1 seek to 'pass on' data—that is, to be both driven by data capture and to be predictive—unforeseen events become critical. Rather than being filtered as noise or dismissed as accidental, they drive system updates. Each crisis is the motor and the end of control systems; each initially singular event is carefully saved, analyzed, and codified. Most succinctly, crises are both what network analytics seek to eliminate and what they perpetuate. In a networked world, there are two operational modes: habitual/programmed repetition (machinic and human) and critical exception.

By grappling with the logic of the update, this chapter seeks to understand how **networks, which are made out of time (memory and real time), threaten to make us out of time.** It begins by outlining how this 'out of time' curiously grounds new media's promise. As an interruption of habit, crises have been crucial to the emergence of new media as 'new.' Starting from an analysis of the rhetorical and theoretical constructions of the Internet as critical, this chapter exposes crisis as new media's critical difference: its norm and its exception. Crises cut through the constant stream of information, differentiating the temporally and temporarily valuable from the mundane, offering its users a taste of real-time responsibility and empowerment. This logic of crisis, however, depends on what seem to be its opposites: codes and habits. Habit supports a worldview driven by automation and automatic codes, which reduces the future to the past, or more precisely, a past anticipation of the future. Rather than being antithetical, though, codes/habits and crises together produce (the illusion of) mythical and mystical human and machinic sovereign subjects who weld together word with action, norm with reality. Exceptional crises justify states of exception that undo the traditional democratic separation of the executive branch from the legislative one.[3] Correspondingly, code is *logos*: code as source, is code conflated with, and substituting for, action.[4] A habit, understood as coded program, is both the source of the action and the action itself.[5] Within both crisis and habit, execution and legislation coincide—programs are all about states of exception.

This twinning of crisis and code/habit has not diminished crises, but rather proliferated them through an unending series of decisions and unforeseen consequences that undermine the agency they promise. From financial crises linked to complex software programs to diagnoses and predictions of global climate change that depend on the use of supercomputers, from undetected computer viruses to bombings at securitized airports, we are increasingly called on both to trust coded systems and to prepare for events that elude them. To displace this twinning, this chapter argues for a practice of exhausting exhaustion: a recovery of the undead potential of our decisions and our information through a practice of constant care. This practice entails a more rigorous engagement with habit, rather than a running away from it. It thus concludes by engaging current theories of involuntary memory in neuroscience, which trouble the notion of habit as something "in memory," for, based on these theories, there is arguably no involuntary memory separate from perception. Every involuntary repetition trains perception, calling into question the notion of memory as an entity we recall. By engaging the unfolding present, we can displace this oscillation between memory and crises, and the dreams of preemption it undergirds.

Internet Critical

The Internet, in many ways, has been theorized, sold, and sometimes experienced as 'critical.' In the mid to late 1990s, when the Internet first emerged as a mass personalized medium through its privatization, both its detractors and supporters promoted it as a "turning point," "a vitally important or decisive state" in civilization, democracy, capitalism, and globalization.[6] Bill Gates called the Internet a medium for "friction-free capitalism."[7] John Perry Barlow infamously declared cyberspace an ideal space outside physical coercion, writing, "governments of the Industrial World, you weary giants of flesh and steel, I come from Cyberspace, the new home of Mind. On behalf of the future, I ask you of the past to leave us alone. You are not welcome among us. You have no sovereignty where we gather."[8] We in cyberspace, he continues, are "creating a world that all may enter without privilege or prejudice accorded by race, economic power, military force, or station of birth. We are creating a world where anyone, anywhere may express his or her beliefs, no matter how singular, without fear of being coerced into silence or conformity."[9] Blatantly disregarding then-current Internet demographics, corporations similarly touted the Internet as the great racial and global equalizer: MCI advertised the Internet as a race-free utopia; Cisco Systems similarly ran television advertisements featuring people from around the world, allegedly already online, who asked viewers, "Are you ready? We are." The phrase "we are" made clear the threat behind these seeming celebrations: YOU should get online because these people already are.[10]

The Internet was also framed as quite literally enabling the critical—understood as enlightened, rational debate—to emerge. Then-U.S. Vice President Al Gore argued that the Global Information Structure finally realized the Athenian public sphere; the U.S. Supreme Court explained that the Internet proved the validity of the U.S. judicial concept of a marketplace of ideas.[11] The Internet, that is, finally instantiated the Enlightenment and its critical dream by allowing people, as Kant prescribed, to break free from tutelage and to express their ideas as writers before the scholarly world.[12] Suddenly, users could all be Martin Luthers or town criers, speaking the truth to power and proclaiming how not to be governed like that.[13] It also, remarkably, instantiated critiques of this Enlightenment dream: many theorists portrayed it as Barthes's, Derrida's, and Foucault's theories come true.[14] The Internet was critical because it fulfilled various theoretical dreams.

This rhetoric of the Internet as critical, which helped transform the Internet from a mainly academic and military communications network into a global medium, is still with us today, even if the daily experience of using the Internet has not lived up to the hype. Irrespective of constant updates, being online has not turned out to be so exciting. Yet, from "Twitter revolutions" to *Facebook.com*'s alleged role in the 2011 protests in Tunisia and Egypt, from hactivism to outraged reactions to Edward Snowden's revelations, Internet technologies are still viewed as inherently linked to freedom. As the controversy over Snowden's revelations also makes clear, this freedom is sometimes framed as calling the critical—and our safety/ security—into crisis.

This crisis is not new or belated: the first attempt by the U.S. government to regulate the content of the Internet coincided with its deregulation. The same government promoting the Information Superhighway condemned it as threatening the sanctity and safety of the home by putting a porn shop in 'our' children's bedroom.[15] Similarly, Godwin's law, which states that "as an online discussion grows longer, the probability of a comparison involving Nazis or Hitler approaches 1," was formulated in the 1990s.[16] So, at the very same time that the Internet (as Usenet) was being trumpeted as the ideal marketplace of ideas, it was also indicted for reducing public debate to a string of nasty accusations. Further, the same corporations celebrating the Internet as the great racial equalizer also funded roundtables on the digital divide.[17] More recently, as I discuss in the next two chapters, the Internet has been linked to cyberbullying and has been formulated as the exact opposite of Barlow's dream: a nationalist machine that spreads rumors and lies. Joshua Kurlantzick, an adjunct fellow at the Pacific Council on International Policy in the United States, told the *Korea Times* in response to the 2008 South Korean beef protests, "the Internet has fostered the spread of nationalism because it allows people to pick up historical trends, and talk about them, with little verification."[18]

Likewise, critics have postulated the Internet as the end of critical theory, not because it literalizes critical theory, but because it makes criticism impossible. As theorists McKenzie Wark and Geert Lovink have argued, the sheer speed of telecommunications undermines the time needed for scholarly contemplation.[19] Scholarship, Wark argues, "assumes a certain kind of time within which the scholarly enterprise can unfold," a time denied by global media events that happen and disappear at the speed of light.[20] Theory's temporality is traditionally belated. Theory stems from the Greek *theoria*, a term that described a group of officials whose formal witnessing of an event ensured its official recognition. To follow and extend Wark's and

Lovink's logic, theory is impossible because we have no time to register events, and we lack a credible authority to legitimate the past as past. In response, Lovink has argued for a "theory on the run" and Wark has contended that theory itself must travel along the same vectors as media events. I am, as I've stated elsewhere, sympathetic to these calls.[21] However, I also think we need to theorize this narrative of theory in crisis, which resonates both with the general proliferation of crises discussed earlier and with recent handwringing within academia over the alleged death of theory. Moreover, we need to theorize this narrative in relation to its corollary: an ever-increasing desire for crises. Theory has embraced crisis, and there are more "turns" in theory now than ever before: from the affective to the nonhuman. Further, there is an unrelenting stream of updates that demand response, from ever-updating *Twitter.com* feeds to exploding inboxes. The lack of time to respond, brought about by the inhumanly clocked time of our computers that renders the new old, coupled with the demand for response, makes the Internet compelling. Crises structure new media temporality. If, as Ursula Frohne theorized in response to the spread of webcams, "to be is to be seen," in the era of social media, "**to be is to be updated.**"[22] **Automatically recognized changes of status have moved from surveillance to evidence of one's ongoing existence.** A true sign of trouble is the lack of signs or updates.

What is therefore remarkable is not that the criticality or newness of new media is called into question but rather that new media continually resurge and spread, even as they disappoint. Soon after the dotcom meltdown—another critical new media moment, which seemed to threaten new media's continued existence—social media emerged as the new promise; *Facebook.com* was heralded as the new Silicon Valley hero. Web 2.0 promised a new beginning, a new future, albeit one not as bold as the original web, which needed no qualifiers. Web 2.0 also promised that the web would never truly die, for every downturn could be treated as temporary, as a call for an update to revamp and renew: Web 3.0, 4.0, ...

This twinned ennui and excitement is not a reaction to new media, it is produced by the very concept of new media. This anticipation coupled with knowing disappointment drives new media's ephemerality and endurance. Hence the point is not to decide whether the Internet is good or bad, but rather to understand how the Internet survives in contradictory yet 'critical' forms: from the Internet as both ruining the economy (dot-bombs) and saving it (dotcoms, Web 2.0) to the Internet as both proliferating democracy (so-called Twitter revolutions) and undermining it (PRISM).

Crisis, New Media's Critical Difference

Crisis is new media's critical difference. In new media, crisis has found its medium; and in crisis, new media has found its value, its punctuating device. Crises have been central to making the Internet a mass medium to end mass media: a ~~mass~~ personalized device. The aforementioned crises answered the early questions: Why go online? And how can the Internet—an asynchronous medium of communication—provide compelling events for users? Further, crises are central to experiences of new media agency, to information as power. Crises—moments that demand real-time responses—make new media valuable and empowering by tying certain information to decisions, personal or political (in this sense, new media also personalize crises). Crises mark the difference between "using" and other modes of media spectatorship/viewing—in particular, "watching" television, which has been theorized in terms of "liveness" and catastrophe. Indeed, the disparity between new media crises and televisual catastrophes encapsulates the promise and threat of new media.

Traditionally, television has most frequently been theorized in terms of "liveness": a constant flowing connection. As Jane Feuer has influentially argued, despite the fact that much television programming is taped, (broadcast) television is promoted as essentially live, as offering a direct connection to an unfolding reality "out there."[23] As Mary Ann Doane has further developed in her canonical "Information, Crisis, Catastrophe," this feeling of direct connection is greatly enhanced in moments of catastrophe: during them, we stop simply glancing at the steady stream of information on the television set and sit transfixed before it. Distinguishing between television's three different modes of framing the event—information (the steady stream of regular news), crisis (a condensation of time that demands a decision: for this reason, it is usually intertwined with political events), and catastrophe (immediate "subjectless" events involving death and the failure of technology)—Doane argues that commercial television privileges catastrophe because catastrophe "corroborates television's access to the momentary, the discontinuous, the real."[24] Catastrophe, that is, underscores television's greatest technological power: "its ability to be there—both on the scene and in your living room. ... [T]he death associated with catastrophe ensures that television is felt as an immediate collision with the real in all its intractability—bodies in crisis, technology gone awry."[25] Rather than showing a series of decisions (or significations), televisual catastrophe presents us with a series of events that promise reference: a possibility of touching the real. However, as in Feuer's critique of 'liveness,' Doane points out that television's relation to catastrophe is ideological rather than essential. Commercial television privileges catastrophes because they make television programming and the selling of viewers' time seem accidental rather than necessary—this programming, after all, can always allegedly be interrupted in case of a catastrophe. Thus, television renders economic crises, which threaten to reveal the capitalist structure that undergirds commercial television's survival, into catastrophes: apolitical events that simply happen.[26]

In contrast, new media are crisis machines: the difference between empowered user and the couch potato, the difference between crisis and catastrophe—that is, different habits of 'touching the real.' From the endless text messages that have replaced the simple act of making a dinner date to the familiar genre of "email forwarding disasters," crises promise to move users from the banal to the crucial by offering the experience of something like responsibility; something like the consequences and joys of 'being in touch.' Crises promise to take users out of normal time, not by referencing the real but rather by indexing real time, by touching times that touch a real, different time: times of real decision; times of their lives. Crises touch duration; they compress asynchronous time. They point to a time that seems to show that our machines can be interrupted, that computer programs can be altered, aborted, or halted in response to crises such as stock market crashes. Further, crises, like televisual catastrophes, punctuate the constant stream of information, so that some information, however briefly, becomes (in)valuable. This value is not necessarily inherent to the material itself: this information could, at other moments, be incidental; but it becomes significant because it relates to an ongoing decision, to a 'real-time' action.

The concept of "real time" has been central to the makeover of computers from work devices into media machines that cut across work and leisure. "Real-time" operating systems transform the computer from a machine run by human operators in batch mode to 'alive' personal machines that respond to users' commands. "Real-time" content—stock quotes, news feeds, and streaming video—similarly transform personal computers into personal media machines. What is real is what unfolds in "real time."[27] If earlier visual indexicality guaranteed authenticity (a photograph was real because it indexed something out there), now "real time" does so, for "real time" points elsewhere: to "real-world" events, to the user's captured actions. That is, "real time" introduces indexicality to this seemingly anti-indexical medium; an indexicality that is felt most acutely in moments of crises, which enable connection and demand response. Crises amplify what Tara McPherson has called "volitional mobility": dynamic changes to web pages in real time, seemingly at the bequest of the user's desires or inputs, that create a sense of "liveness on demand."[28] Volitional mobility, like tele-visual 'liveness,' produces continuity, a fluid path over discontinuity.[29] It is a simulated mobility that expands to fill all time, but simultaneously prom-ises that we are not wasting time, that indeed, through "real time," we touch real time.

The decisions we make, however, seem to prolong crises rather than end them, trapping us in a never-advancing present. Consider, for instance, 'viral' email warnings about viruses. Years after computer security programs had effectively inoculated systems against a 2005 Trojan attached to a mes-sage claiming that Osama bin Laden had been captured, messages about the virus—many of which exaggerated its power—still circulated.[30] These mes-sages spread more effectively than the viruses they warn of. Out of good will, users disseminate these warnings to those in their address books, and then forward warnings about these warnings, etc., etc. (Early on, trolls took advantage of this temporality, through volleys that unleashed a firestorm of warnings against feeding the troll.) These messages, in other words, act as retroviruses. Retroviruses, such as HIV, are composed of RNA strands that use the cell's copying mechanisms to insert DNA versions of themselves into a cell's genome. Similarly, these fleeting messages survive by users copying and saving them, by their active incorporation of these warnings into ever-repeating archives. Through users' efforts to foster safety, they spread retrovirally, flooding the Internet and defeating a computer's antivi-ral systems.

This voluntary yet never-ending spread of information seemingly belies the myth of the Internet as a "small world." As computer scientists D. Liben-Nowell and J. Kleinberg have shown, the spread of chain letters resembles a long thin tree, rather than a short fat one (see figure 2.1).[31] This diagram seems counterintuitive, for if everyone on the Internet was really within six degrees of each other, information on the Internet should spread quickly and then die. Liben-Nowell and Kleinberg pinpoint asynchrony and dissimilar replying preferences as the cause: because everyone does not forward the same message at once or to the same number of people, mes-sages circulate at different paces and never seem to reach an end.

This temporality—this long thin chain of transmission—seems to describe more than just the spread of chain letters. Consider, for instance, the ways in which a simple search can lead to hours of tangential surfing. Microsoft playfully called this temporality "search overload syndrome" in its 2009 advertisements to launch its "decision engine," *Bing*. In these com-mercials, characters responded to a simple query, such as "we really need to find a new place to go for breakfast," with a long stream of unproductive

Figure 2.1
From Liben-Nowell and Kleinberg, "Tracing Information Flow on a Global Scale Using Internet Chain-Letter Data," page 4635. © 2008 National Academy of Sciences, U.S.A.

associations, such as statistics about "The Breakfast Club." These characters were unable to respond to a suggestion—to make a decision—because each word unleashed a long thin chain of references due to the inscription of information into "memory"; they became zombies whose minds were hijacked by involuntary associations. Habitual crises make information persist.

This situation encapsulates Lauren Berlant's description of the present as a cul-de-sac, an impasse. As she explains, the impasse "is a stretch of time in which one moves around with a sense that the world is at once intensely present and enigmatic, such that the activity of living demands both a wandering absorptive awareness and a hypervigilance that collects material that might help to clarify things, maintain one's sea legs, and coordinate the standard melodramatic crises with those processes that have not yet found their genre of event."[32] The present as "crisis ordinary" makes survival, however detrimental to long-term success, an accomplishment: we live in a moment of cruel optimism, in which an attachment to the promises of the good life is precisely what allows one to tread water, but not to swim.[33]

The belief in memory as storage, combined with the belief in "real time" as indexical, is a form of cruel optimism: memory, which once promised to save users from time, makes them out of time by making them respond constantly to information they have already responded to, to things that will not disappear. Information is curiously undead, constantly regenerating, and users save things, if they do, by making the ephemeral endure. As noted in the previous chapter, users save things digitally, if they do, by making what is stable ephemeral. They perversely take what is more lasting—what can remain and still be read for a long duration, such as paper—and make it more volatile. This digital 'version' is more volatile not simply because magnetically stored data decay more quickly than paper, but also because software and hardware constantly change, manically upgrade. This relentlessly upgrading system ensures that things that simply remain digitally cannot easily be read, for we (some combination of ourselves and our machines) must constantly migrate, regenerate—that is, care for now—anything that we want to remain to be cared for and read.

As endless searches reveal, the sheer amount of saved (that is, constantly regenerating) information seems to defer the future it once promised. Memory, which was initially posited as a way to save users by catching what they lose in real time, by making the ephemeral endure and thereby fulfilling that impossible promise of history to gather everything into the present, now threatens their sanity; but only if they expect engines and information to make their decisions for them, only if they expect their programs to (dis)solve their real-time crises, only if they accept the premise of automation that grounds the conversion of accidents into crises.

Bing's promised solution—the exhausting of decisions altogether through a "decision engine" (which resonates with calls for states of emergency to exhaust crises)—is hardly empowering. *Bing*'s promised automation, however, does inadvertently reveal that, if "real-time" new media do enable user agency, they do so in ways that mimic, rather than belie, automation and machines. Machinic "real time" and crises are both decision-making processes. According to the *OED*, "real time" is "the actual time during which a process or event occurs, especially one analyzed by a computer, in contrast to time subsequent to it when computer processing may be done, a recording replayed, or the like."[34] Crucially, hard and soft real-time systems are subject to a "real-time constraint." That is, they need to respond, in a forced duration, to actions predefined—captured—as events. In computer systems, "real time" reacts to the live: their 'liveness' is their quick acknowledgment of and response to users' actions. Computers are "feedback machines," based on control mechanisms that automate decision making. As the definition of "real time" makes clear, "real time" refers to the time of computer processing, not to the user's time. "Real time" is never real time—it is deferred and mediated. The emphasis on crises as central to user agency screens the ever-increasing automation of their decisions. While users struggle to respond to "What's on your mind?," their machines quietly disseminate their activity. What they experience is arguably not a real decision, but rather the already planned in an unforeseen manner: increasingly, user decisions are like actions in a video game. They are immediately felt, affective, and based on user actions, and yet at the same time programmed. Furthermore, crises do not arguably interrupt programming, for crises—exceptions that demand a suspension, or at the very least an interruption of rules or the creation of new norms—are intriguingly linked to technical codes or programs, that is, machine and human habits.

Logos as State of Exception

Crises, and the decisions they demand, do not simply lead to experiences of responsibility; as the term "panic button" nicely highlights, they can also produce moments of fear and terror from which people want corporate, governmental, or technological intermediaries to save them.[35] States of exception are now common reactions to unforeseen events that call for extraordinary responses, to moments of what Jacques Derrida has called undecidability. According to Derrida, the undecidable calls for a response that "though foreign and heterogeneous to the order of the calculable and the rule, must ... nonetheless ... deliver itself over to the impossible decision while taking account of law and rules."[36] States of emergency respond to the undecidable by constructing a sovereign subject who closes the gap between rules and decisions, who knits together force and law (or, more properly, force and suspended law); this sovereign subject, through his actions, makes the spirit of the (dead) law live. Although these states seem to be the dead (that is, living) opposite of codes and programs, they are linked together through questions of agency or, more properly, as I explain later, authority.

Giorgio Agamben has most influentially theorized states of exception. He notes that one of the essential characteristics of the state of exception is "the provisional abolition of the distinction among legislative, executive, and judicial powers."[37] This provisional granting of "full powers" to the executive suspends a norm, such as the constitution, in order to better apply it. Agamben describes the state of exception, which responds to a crisis that challenges the efficacy of norms, as

the opening of a space in which application and norm reveal their separation and a pure force-of-~~law~~ realizes (that is, applies by ceasing to apply ...) a norm whose application has been suspended. In this way, the impossible task of welding norm and reality together, and thereby constituting the normal sphere, is carried out in the form of the exception, that is to say, by presupposing their nexus. This means that in order to apply a norm it is ultimately necessary to suspend its application, to produce an exception. In every case, the state of exception marks a threshold at which logic and praxis blur with each other and a pure violence without *logos* claims to realize an enunciation without any real reference.[38]

The state of exception thus reveals that norm and reality are usually separate—it responds to the moment of their greatest separation. In order to bring them together, force without law or *logos*—a living sovereign—authorizes a norm "without any reference to reality."[39] It is a moment of pure violence without *logos*. That is, if the relationship between law and justice—a judicial decision—usually refers to an actual case (it is an instance of parole, an act of speaking), a state of exception is langue in its pure state: language in the abstract and at its most mystical.

Given this, states of exception would seem the opposite of programmed and habitual actions. In neuroscience, habits—as one more turn in the ever twisted and twisting double helix that binds together biology and computation—are described as chunked programs "in memory."[40] Habits, like programs, are forms of automaticity: processes linking certain behaviors to cues that are central to human prediction and function. Unlike most computer programs (that is, outside of machine learning algorithms), habits develop slowly over time. However, once established, they are "strongly and reliably activated," regardless of any initial rewards or goals.[41] Habits are also described as "inflexible." Although goals can be satisfied in various ways, there is only one way to satisfy a habit: by repeating it exactly.[42] The relative autonomy and inflexibility of habit is explained through the concept of memory "chunking," where a "chunk is an integrated memory representation that can be selected as a whole and executed with minimal attentional involvement."[43] Researchers have deduced the existence of these chunks via neural mapping, which reveals a spike in activity at the beginning and end points of a habitual action sequence and minimal response in the middle. Habits are formed slowly, it is thought, in order to ensure that only truly important actions become "chunked" and thus made relatively autonomous, automatic, and inflexible. Importantly, habit learning is nonconscious. As Ann Graybiel has argued in her influential article "The Basal Ganglia and Chunking of Action Repertoires," habit learning in humans "is characterized by two key features: lack of awareness of the algorithm learned, and a slow rate of acquisition."[44] Habits are trained algorithms, stored in involuntary memory.

The habit "algorithm" is intriguingly both the source of action and the action itself, the source of learning and what is learned. As Graybiel puts it: a "paradox in the learning literature is that the basal ganglia are at once thought to be important for the production of habitual or 'automatic' responses and yet are thought to be important for new S–R learning. Does the same mechanism subserve both learning and expression of habits?"[45] Graybiel resolves this paradox by arguing for the basal ganglia as central to the construction of performance units (and other models hypothesize two pathways connecting sensory association areas to the premotor cortex, thus making the basal ganglia loop central to the training of direct corticocortical pathways, rather than the place where habitual behaviors are stored).[46] However, habit, as Clare Carlisle points out, is oddly both source (or neurocomputationally speaking, algorithm) and action: I perform a habit out of habit.[47] Even further, because I perform a habit, I must like it. Neuroscience views the coincidence between human volition and habits as "misattribution": "people may misattribute externally-cued representations to their own natural response to the situation, that is, to their internal preferences and desires."[48] Because they do something often, they may reason that they like it; because they do something often, they believe their feelings must be responsible. This oddly reversed temporality—because I do it, I must like it; because I do it, I must be the source—resonates with the very work of habit, namely prediction. Graybiel, in terms that strongly echo David Hume's and Henri Bergson's discussions of habit, states, "prediction should be a crucial guide to the implementation of chunks and to their formation. ... Prediction would allow the initiation of chunks and their formation."[49] (Parkinson's disease with its jerky motions, Graybiel hypothesizes, stems from the inability of patients to properly predict actions.) Habits would then become "bad," or problems that need to be "updated," when they become faulty predictors—when they lead to incorrect anticipations due to changes in the environment—or when the actions they invoke conflict with current goals.

Habit is both the source and the action because it is 'programmed.' The logic of programming reduces the living world to dead writing; programs condense everything to "source code" written in advance, hence the adjective "source."[50] This is because code allegedly does what it *says*.[51] This understanding presumes no difference between source code and execution, instruction and result. It perversely renders code, because of machinic, dead repetition, into *logos*. Like the King's speech in Plato's *Phaedrus*, code does not pronounce knowledge or demonstrate it; it transparently pronounces itself, for it does what it says.[52] The hidden signified—meaning; the father's intentions—shines through and transforms itself into action. Like Faust's translation of *logos* as "deed"—"The spirit speaks! I see how it must read / And boldly write: 'In the beginning was the Deed!'"—software is word become action: a replacement of process with inscription that makes writing a live power by conflating force and law.[53] By converting action into language, source code emerges. To put it slightly differently and to revise an example from *Programmed Visions*, the fact that lawyers working at the intersections of new media and law declare that "code is law"—that programs encode certain regulations—is, at one level, hardly profound.[54] Code, after all, is "a systematic collection or digest of the laws of a country, or of those relating to a particular subject";[55] but "code as law" makes code something different: automatically executable. This executability makes code not law but rather every lawyer's dream of what law should be: automatically enabling and disabling certain actions, and functioning at the level of everyday practice. **Code as law is code as police.** Insightfully, Derrida argues that modern technologies ensure the sphere of "the police absolute ubiquity."[56] The police weld together norm with reality; they "are present or represented everywhere there is force of law. ... They are present, sometimes invisible but always effective, wherever there is preservation of the social order."[57]

Code as law as police, like the state of exception, makes executive, legislative, and juridical powers coincide. Code as law as police erases the gap between force and writing, langue and parole, in a complementary if reverse fashion to the state of exception. It makes language abstract—erases the importance of enunciation—not by denying law, but rather by making *logos* everything. Code is executable because it embodies the power of the executive. More generally, the dream of executive power as source lies at the heart of Austin-inspired understandings of performative utterances as simply doing what they say. As Judith Butler has argued in *Excitable Speech*, this theorization posits the speaker as "the judge or some other representative of the law."[58] Code resuscitates fantasies of sovereign, or executive, structures of power. It embodies "a wish to return to a simpler and more reassuring map of power, one in which the assumption of sovereignty remains secure."[59] Not accidentally, programming in a higher-level language has been compared to entering a magical world: a world of *logos*, in which one's code faithfully represents one's intentions, albeit through its blind repetition rather than its "living" status.[60] As Joseph Weizenbaum, MIT professor, creator of ELIZA, and member of the famed MIT AI lab, has argued:

The computer programmer ... is a creator of universes for which he alone is the lawgiver. So, of course, is the designer of any game. But universes of virtually unlimited complexity can be created in the form of computer programs. Moreover, and this is a crucial point, systems so formulated and elaborated *act out* their programmed scripts. They compliantly obey their laws and vividly exhibit their obedient behavior. No playwright, no stage director, no emperor, however powerful, has ever exercised such absolute authority to arrange a stage or a field of battle and to command such unswervingly dutiful actors or troops.[61]

Weizenbaum's description underscores the mystical power at the base of programming: a power both to found and to enforce. Automatic compliance welds together script and force, again, code as law as police or as the end of democracy. As Derrida has underscored, the police is the name for "the degeneration of democratic *power*. ... Why? In absolute monarchy, legislative and executive powers are united. In it violence is therefore normal, conforming to its essence, its idea, its spirit. In democracy, on the contrary, violence is no longer accorded nor granted to the spirit of the police. Because of the presumed separation of powers, it is exercised illegitimately, especially when instead of enforcing the law, it makes the law."[62] Code as *logos* and states of exception both signify a decay of the decay that is democracy.

Tellingly, this machinic execution of law is linked to the emergence of a sovereign user. Celebrations of an all-powerful user/agent—YOU as the network, YOU as "produser"—counteract concerns over code as law as police by positing YOU as the sovereign subject, YOU as the decider. An agent, however, only holds power through proxy; an agent acts on behalf of another subject. In networks, the real power would seem to be technology, rather than the users or programmers who authorize actions through their commands and clicks. Programmers and users are not creators of languages, nor the actual executors, but rather living sources that take credit for the action. Similarly, states of exception rely on *auctoritas*. The *auctor*, Agamben explains, is one who, like a father, 'naturally' embodies authority and authorizes a state of emergency.[63] An *auctor* is "the person who augments, increases or perfects the act—or the legal situation—of someone else."[64] The subject that arises, then, is the opposite of the democratic agent, whose power stems from *potestas*. Hence the state of exception, Agamben argues, revives the *auctoritas* as father, as living law:

The state of exception … is founded on the essential fiction according to which anomie (in the form of *auctoritas*, living law, or the force of law) is still related to the juridical order and the power to suspend the norm has an immediate hold on life. As long as the two elements remain correlated yet conceptually, temporally, and subjectively distinct (as in republican Rome's contrast between the Senate and the people, or in medieval Europe's contrast between spiritual and temporal powers) their dialectic—though founded on a fiction—can nevertheless function in some way. But when they tend to coincide in a single person, when the state of exception, in which they are bound and blurred together, becomes the rule, then the juridico-political system transforms itself into a killing machine.[65]

The reference here to killing machines is not accidental. States of exception make possible a living authority based on an unliving (or, as my spell checker keeps insisting, an unloving) execution. This insistence on life also reveals why all those discussions of code anthropomorphize it, using terms such as "says" or "wants." It is, after all, as a living power that code can authorize. It is the father behind *logos* that shines through the code.

To summarize, we are witnessing an odd dovetailing of the force of law without law—the state of exception—with writing as *logos*. This perverts the perversion that writing was supposed to be (writing as the bastard "mere repetition" was defined in contrast to *logos*). They are both language at its most abstract and mystical, albeit for seemingly diametrically opposed reasons, for one is allegedly language without writing; the other writing without language. This convergence, which is really a complementary pairing since they come to the same point from different ends, puts in place an originary sovereign subject. This originary sovereign subject, however, as much as she or he may seem to authorize and begin the state of exception, is created belatedly by it. Derrida calls sovereign violence the naming of oneself as sovereign: the sovereign "names itself. Sovereign is the violent power of this originary appellation"—an appellation that is also an iteration.[66] Butler similarly argues that through iterability, the performative utterance creates the person who speaks it. Further, the effect of this utterance does not originate with the speaker, but rather with the community she or he joins through speaking.[67] The programmer/user is produced through the act of programming. As I've argued extensively in *Programmed Visions*, code as *logos* depends on many circumstances, which also undermine the authority of those who would write.

Habit + Crisis = Update

To counter this sovereignty that is no sovereignty, we need to take seriously the dynamic processes that make and also undermine code or habit *logos*. Habit needs to be rethought in relation to memory, in terms of repetition and difference. Habits, for all their inflexibility, are also defined by their changeability: they, unlike other involuntary actions, can be changed and result from change; they are "second nature." Intriguingly, almost all the literature, from the psychological to the philosophical, assumes that consciousness drives habit change and, further, that conflict/crisis drives this consciousness. Wood et al., for instance, argue that people usually only become aware of habits when their habits conflict with their current goals and intentions; Charles Duhigg writes, "in the heat of a crisis, the right habits emerge";[68] Burkitt via Bourdieu asserts that reason and criticism arise in situations in which there is a conflict of customs within the *habitus*, revealing the need for the reorganization and reconstruction of social institutions.[69]

Crises are opportunities for habit change, precisely because they disrupt contexts and undermine the efficacy of habitual anticipations. Crises frame habits—as "vestige[s] of past goal pursuit"—as what must be changed in order for the present to once again become a past that can operate for the future.[70] It should be no surprise, then, that the proliferation of crises have been so central to neoliberalism, as proselytized by Milton Friedman and as criticized by Naomi Klein.[71] Again, the impasse, or the thin, never-ending chain of decisions, embodies this proliferation of crises: an affectively intense present that goes nowhere. **The constant update, that is, deprives habit of its ability to habituate.** As soon as one is comfortable, habits are disrupted, so that one is always dependent on and responsive to the environment. One must constantly respond in order to remain close to the same.

Perhaps, but to decouple habit from crisis—to make habit once again habituate—we need to revisit habit and code as automatic execution. The conflation of code with execution relies not only on code as *logos*, but also on memory as simple recall.[72] It assumes that memory is storage and can be localized in an organ. Recent developments in neurobiology focused on the automatic nature of the nervous system—or, to use John von Neumann's terms, the complex human automaton—show, however, that there is no single memory organ: there is no one organ, separate from other organs/cellular circuits, that simply stores information.[73]

This is a point made elliptically in Nobel Prize recipient Eric R. Kandel's book *In Search of Memory*, which combines autobiography, history, and science to relay advances in neurobiology to a lay public.[74] As Kandel explains, memory is now divided into explicit and implicit memory, where explicit memory is linked with consciousness (with the conscious recall of people, places, objects, facts, and events) and implicit memory (unconscious, procedural memory) underlies classic habitual behavior and its mechanisms for change. Constant repetition can transform explicit memory into implicit memory (and vice versa). Implicit memory underlies habituation (the acclimation of organisms to certain types of signals), sensitization (the opposite of habituation; hence, enhanced alertness to usually noxious signals), and classical Pavlovian conditioning (the coupling of an innocuous with a noxious signal, so that an animal reacts to the benign signal as though it were dangerous). Implicit memory is intimately tied to perception and motor skills, to nonconscious, mechanical, constantly repeated and reflexive actions; it makes humans automata (a comparison that von Neumann would engage in more deeply in his later years). Implicit memory refers, as K. B. McDermott argues, to manifestations of memory that occur in the absence of intentions to recollect.[75] **It is knowledge without knowing.** It is memory that is directly performed without any awareness of memory.[76]

The classic distinction between these two types of memory, as Kandel explains, stems from the most famous case of anterograde amnesia: H.M. Patient H.M. (later revealed to be Henry Gustave Molaison) underwent surgery to cure his epilepsy in 1953. During this surgery, H.M.'s neurosurgeon, Dr. William Scoville, removed the inner surface of his medial temporal lobes and hippocampus from both sides of his brain. After the surgery, H.M. could not make any new long-term explicit memories. He could remember things before the surgery and his short-term working memory was functional, but he could not convert short-term memories into long-term ones. So, although he would work with the neuroscientist Brenda Milner for many years, he would greet her each time as though for the first time.

H.M. could however, as Milner discovered, still learn things, still acquire new habits, albeit without realizing it. Famously, she taught him how to draw each day and—even as he could never remember learning to draw—he became better at drawing every day. H.M.'s case was thus key to disproving Karl Lashley's notion that memories were not localized, but also to proving that there were two physically separate memory systems—implicit and explicit memory—and that there were two types of memory within these memories—short-term and long-term. As Kandel puts it, Milner proved that Freud's notion of unconscious memory was correct, for most of our actions are unconscious.[77]

But what is implicit memory? Kandel's experiments with *Aplysia*, a type of snail whose cultured neurons he 'trained' in various ways (classic habituation, sensitization, and classic conditioning), have been key to understanding the neurobiological underpinnings of implicit memory. In these experiments, sensitization became the strengthening of a tie and habituation the weakening of one. (These experiments, of course, assume that *Aplysia* have only implicit memory.)

Using cultured and isolated *Aplysia* neurons, Kandel was able to show that in short-term implicit memory, the connection—that is, the amount of neurotransmitter glutamate—between the sensory and motor neuron is strengthened.[78] In long-term memory, more ties—at a specific synapse—are created.[79] Importantly, the maintenance of this new connection does not happen at the nucleus, but rather at the axon terminal itself. The intense serotonin pulses convert the prion-like protein CREB into the dominant form (a prion is a protein that can fold into two distinct shapes, one of which is dominant and the other recessive). In the dominant form, the protein is self-perpetuating: it causes the recessive form to change its shape and become dominant and self-perpetuating. This dominant form activates the dormant messenger RNA, which regulates the synthesis at the new synaptic terminal and stabilizes the synapse.

The actions of prions reveal that one constantly repeats in order to remain close to the same. Hence the seeming permanence of memory, a seeming permanence that relies on a constant regeneration: a process as destructive as it is as constructive. Mad cow disease, caused by a prion, leads to mad humans: humans who cannot remember or recall. The same mechanism lies at the heart of both the destruction and construction of long-term memory.

In vertebrates, implicit memory is more complex and involves the amygdala; it involves changes to the very sensory system that initially perceived the event. In memory recall, the "neurons that retrieve the memory of the stimulus are the same sensory and motor neurons that were activated in the first place."[80] Importantly, for every action, there are two pathways that are involved: one that is direct (implicit) and another that goes through the cortex (explicit). The basic idea is the same as in invertebrates: training leads to a strengthening of the connections between the neurons. Kandel divides the difference between implicit and explicit memory into bottom-up (serotonin-based) changes and top-down (dopamine-based) responses, where dopamine is central to the stabilization of the "spatial map" in the hippocampus. The outside world provokes involuntary memory; in contrast, voluntary attention arises from the "internal need to process stimuli that are not automatically salient."[81] This would imply that habitual memory (to the extent that habit and memory can be separated) is provoked externally, whereas conscious recall is provided internally.[82]

We still do not know everything about memory, especially the relation between explicit and implicit memory, and Kandel's text is not the end of the story. What MRIs really show and the specifics of the biological bases of mental illness are not definitively known.[83] Regardless, this work does pose the question: If memory is the strengthening and development of certain pathways—pathways that are not particular to memory but to sensory motor action and perception—is memory something separate, something that is stored? That is, what does memory have to do with memory? Does memory even exist? Although Kandel does not explicitly raise this question, others working on implicit memory do, arguing that implicit memory is a form of long-term priming that has more to do with pattern recognition and perception than with explicit memory. That is, implicit memory has little to do with meaning, with signification: with the recall of a signified from a signifier, meaning from a sign. Implicit memory—changes that persist—is learning.[84] This kind of learning, further, cannot be localized to the brain. If memory, as E. Tulving has suggested, "has to do with the aftereffects of stimulation at one time that manifest themselves subsequently at another time,"[85] then memory itself, as Henry Roediger III has argued, engages far more than the central nervous system): our immune system, for instance, would also seem to be a memory system.[86] Wood herself argues that habit is not a knowledge structure, but a response pattern and a control mechanism.[87]

Rather than arguing whether memory exists and what it is or is not, what is most needed is a change of perspective, one that acknowledges that memory is an action, an activation and difference in structure, perhaps making memory not anything because it is everything. This change in perspective would make the pertinent questions not "What is memory?" and "Where is memory stored?" but instead "What relations are central to habituation/training, which are evidence of both remembering and forgetting?" To what extent, that is, can we consider memory a habit of living that involves repetition as both living and dead?[88] Kandel links memory to continuity: "Without the binding force of memory, experience would be splintered into as many fragments as there are moments in life. ... Memory is ... essential for the continuity of individual identity."[89] Without memory, or life as a consistent thing, our sense of self as a being disintegrates, but also, "when such changes persist, the result is memory storage."[90] The evidence of memory storage is what it supposedly enables: persistence.

This notion of being as persistence is also central to more philosophical theorizations of habit. Félix Ravaisson most influentially argued that habit is central to a being's tendency to persist: habit signals a change in disposition—a disposition toward change—in a being, which does not change, even as it does.[91] Elizabeth Grosz, drawing from Ravaisson, Henri Bergson, and Gilles Deleuze, argues that habit is a "creative capacity that produces the possibility of stability in a universe in which change is fundamental. ... [It] organize[s] lived regularities, moments of cohesion and repetition, in a universe in which nothing truly repeats."[92] Crucially, for Grosz, "habit not only anchors a site of regularity in a universe of perpetual change; it initiates change in the apparently unchanging, it opens up the possibility of understanding the very force of temporality itself, the force that adheres the past to the present and orients both to the possibilities of action in the future."[93] For Grosz, habit enables radical change through reflection, by offering us the time needed to produce the truly different.

Perhaps, but this repetition, which enables a precarious persistence both within and outside the machine, also calls for responsibility: constant decisions to save and for an uncertain safety. Saving is something that technology or bodies cannot do alone; the battle to save is a crisis in the strongest sense of the word. This necessary repetition makes us realize that our desire for safety as simple securing, as ensured by code, actually puts us at risk of losing what we value, whether it is data stored on old floppy drives or habituations.[94] It also forces us to engage with the fact that if some things stay in place, it is not because they are unchanging and unchangeable, but rather because they are constantly implemented and enforced. From regenerative mercury delay line tubes to habits of living, what remains is not what is static, but instead what constantly repeats and is repeated. This does not mean that there are no things that can be identified later as sources, but rather that constant motion and care recalls things. Further, acknowledging this necessary repetition moves us away from wanting an end (because what ends will end) and toward actively engaging and taking responsibility for everything we want to endure. It underscores the importance of access, and makes us see how digitization can be a means of dynamic preservation. To access repeatedly is to preserve through construction (and sometimes destruction).

This notion of constant care can exhaust the kind of exhaustion encapsulated in "search overload syndrome" and perhaps in the "crisis ordinary." The experience of the undecidable—with both its reliance on and difference from rules—highlights the fact that any responsibility worthy of its name depends on making a decision in the face of uncertainty. As Thomas Keenan eloquently explains, "the only responsibility worthy of the name comes with the removal of grounds, the withdrawal of the rules or the knowledge on which we might rely to make our decisions for us. No grounds means no alibis, no elsewhere to which we might refer the instance of our decision."[95] Derrida similarly argues, "a decision that would not go through the test and ordeal of the undecidable would not be a free decision; it would only be the programmable application or the continuous unfolding of a calculable process."[96] The undecidable is thus freedom in the rigorous sense of the word; a freedom that comes not from security but rather from risk. It is a moment of pause that interrupts our retroviral dissemination and induces the madness that, as Derrida (via Kierkegaard) has argued, accompanies any decision.[97] The madness of a decision, though, differs from the madness described by Microsoft, which stems from the constant deferral of a decision and a desire for a simple answer.[98] To exhaust exhaustion, we need to exhaust too the desire for an end, for a moment in which things just can stand still.

To exhaust exhaustion, we must also deal with, and emphasize, the precariousness of programs, habits, and their predictions. That is, if we use programs and habits to help save the future—to fight the exhaustion of planetary reserves, etc.—we must frame the gap between their predictions and the future as calls for responsibility, rather than as potential errors or truths. 'Trusting' a program does not mean simply accepting its predictions or letting it decide the future, but instead acknowledging the impossibility of knowing its truth in advance while nonetheless responding to its results. Models, in other words, are modes of 'hypothesis,' and we lose the fight if we assume that models are or should be the truth—that they should be 'hyperreal.' To return to Antoinette Rouvroy's discussion of autonomic computing discussed in the previous chapter, the fact that models actualize all statistical probabilities does not have to foreclose the future—it can be the grounds for creating new and different ones.[99]

Global climate models make this clear. These models hope to defer the future. These predictive models are produced so that, if they are persuasive and thus convince us to cut back on our carbon emissions, what they predict will not happen—that is, their predictions will not be verifiable. This relationship is necessary because, by the time we know whether their exact predictions are accurate, it will be too late (this is perhaps why the second Bush administration supported global climate change research: by investigating the problem, building better models, they bought more time for polluters). I stress this temporality because by framing this temporality in terms of responsibility, we can best respond to critics who deride these models by focusing on the fallibility of algorithms and data, as if the gap between the future and future predictions was reason for dismissal rather than hope.[100]

To defer a future for another future is to engage with the undead of information. The undead of information haunts the past and the future; it is itself a haunting. As Derrida explains, "the undecidable remains caught, lodged, as a ghost ... in every decision, in every event of decision. Its ghostliness ... deconstructs from within all assurance of presence, all certainty or all alleged criteriology assuring us of the justice of a decision, in truth of the very event of a decision."[101] This undeadness means that a decision is never decisive, that it can always be revisited and reworked. Repetition is not simply exhaustion: not simply repetition of the same that uses up its object or subject. What can emerge positively from the linking of crises to networks— what must emerge from it, if we are not to exhaust our resources and ourselves—are constant ethical encounters between the self and other. These moments can call forth a new future, a way to exhaust exhaustion, even as they complicate the deconstructive promise of responsibility by threatening a present of affectively draining, yet sustaining crises. Undecidable and undead indeed.

Part II Privately Public: The Internet's Perverse Subjects

On June 5, 2013, Glenn Greenwald published an article in the *Guardian* exposing a secret court order requiring Verizon to give the National Security Agency (NSA) metadata—such as phone numbers called, duration of call, and location—about all telephone calls in its system on an "ongoing, daily basis."[1] This was the first of many leaks which would outline shadowy intergovernmental systems to capture and analyze the metadata of all U.S. and foreign telephone calls (such as the data visualization tool Boundless Informant), as well as to examine the content of foreign telecommunications that traveled through U.S.-owned routers (PRISM)—all without warrants. In addition, the U.S. government was able to access content of civilian data through Tempora, a U.K.-based system that basically downloaded all traffic traveling through the United Kingdom.

Again, what was arguably most shocking about Edward Snowden's revelations is the fact that they counted as revelations. That the NSA regularly downloads traffic from Internet backbones had been known since 2004, after former AT&T technician Mark Klein exposed this practice and tied it to a far-reaching spying program.[2] As well, *Smith v. Maryland*—a case from 1979—held that the U.S. government did not need warrants in order to access telephone numbers that the defendant had called, because he had "no legitimate expectation of privacy" about this information.[3] Further, section 215 of the U.S. Patriot Act granted the director of the Federal Bureau of Investigation or his or her designate the ability to apply for "an order requiring the production of any tangible things (including books, records, papers, documents, and other items) for an investigation to obtain foreign intelligence information not concerning a United States person or to protect against international terrorism or clandestine intelligence activities, provided that such investigation of a United States person is not conducted solely upon the basis of activities protected by the first amendment to the Constitution."[4] Also given that capturing all and then searching later for pertinent information is much easier than selective downloading in real time (search terms are only obvious after an event), the existence of a mass storage system should have surprised no one.

Even more disturbing than the revelation that was no revelation was the corporate outrage that was no outrage. The anger and allegations of betrayal expressed by companies such as Google and Facebook against the U.S. government ring false on multiple levels, for the NSA's information was and is so invaluable and insidious because of back doors, Real Names, and unique identifiers implemented by these corporations.[5] **Surveillance is now a state- and privately funded co-production.** As the next chapter explains, these identifiers and tracking mechanisms—introduced as a way to 'secure' the Internet—have transformed the Internet into a series of poorly gated, trackable communities, which are hardly 'safe' spaces for users. They have, however, made the Internet a viable commercial space: not only a place for market transactions but also itself a market, for the Internet is now Big Data. They have turned once silent and private acts—such as reading a book—into noiselessly noisy ones, eroding the difference between reading, writing, and being written. That is, if once, as D. A. Miller most famously argued, however much readers might identify with protagonists (for example, no matter how much one likes and is like Oliver Twist), there is a fundamental and ontological difference between them (one is not Oliver Twist)—for no matter how much a reader "inclusively sees, he is never seen in turn, invisible both to himself (he is reading a novel) and to others (he is reading it in private)"[6]—now the difference between characters and readers has been eroded. Readers, like literary creatures, are "conspicuously encased yet so transparent they are inside-out"; they are characters in a drama putatively called Big Data.[7]

This fundamental exposure inverts the "traditional" boundary between public and private, epitomized by the window. The window, as Thomas Keenan has demonstrated, grounds liberal political theories of the subject:

The window implies a theory of the human subject as a theory of politics, and the subject's variable status as public or private individual is defined by its position relative to this window. Behind it, in the privacy of home or office, the subject observes that public framed for it by the window's rectangle, looks out and understands prior to passing across the line it marks—the window is this possibility of permeability—into the public. Behind it, the individual is a knowing—that is, seeing, theorizing—subject. In front of it, on the street, for instance, the subject assumes public rights and responsibilities, appears, acts, intervenes in the sphere it shares with other subjects.[8]

Keenan troubles this neat separation of private theorizing subject from actor on the street by asking, "what comes through a window?"[9] How does the light, which makes possible human sight, also undermine the voyeurism it seemingly enables and encourages? The public, Keenan postulates, is glare and intrusion—it is the intrusion of everything other in the self. Further, he stresses, without the glare of publicity, there would be no intimacy, no politics, no subjects.[10] To return to the earlier discussion of habit, **habit is publicity**: it is the experience, the scar, of others that linger in the self. Habits are "remnants" of the past—past goals/selves, past experiences—that live on in our reactions to the environment today, as we anticipate tomorrow. Through habit, we inhabit and are inhabited by alterity.

Although Keenan's discussion of windows focuses on television, it seems to extend to the Internet; but personal computers do not merely retain and disseminate windows, they also explode them. Desktop windows, as Anne Friedberg has shown, are unwindow-like in their layered multiplicity and gravity-defying orientation.[11] Further, these windows are like opaque one-way mirrors, for they reflect text back to the user, and the user cannot look beyond the screen to see who sees it. This unknowable reverse gaze—or its possibility—brings another dimension to the question, "What comes through the window?" As I argue in chapter 4, existing social media platforms reverse the position of public/private, because the subject who acts is increasingly on the inside rather than the outside. **Subjects act publicly in private, or are 'caught' in public acting privately.** This leads to anxiety about privacy and surveillance and to morally outraged attempts to 'fix' this boundary across the political spectrum, from slut shaming to Snowden's leaks. It leads, that is, to what I call an "epistemology of outing": the revelation of mostly open secrets to secure a form of privacy that offers no privacy. (As I discuss in more detail later, this epistemology of outing ruptures the foundational private/public divide described by Eve Sedgwick as the "epistemology of the closet.")[12]

The next two chapters also explore the questions: What if, rather than trying to secure this window by screening its reversal, we refused this framing altogether? What would happen if users warily embraced, rather than hid or were hidden from, the inherently public and promiscuous exchange of information that grounds TCP/IP? What if they focused on creating and inhabiting public spaces online and offline, rather than accepting transparent bubbles of privacy which render them transparent? This next section reveals that such a reframing would move the focus away from protecting or condemning private subjects, who are treated as potential leaks, toward sheltering public subjects, whose actions are currently 'privatized,' that is 'housed' by states and corporations. It would end slut shaming and turn the debate toward the creation and maintenance of public rights. Within the United States, "public figures," a category that includes those involuntarily thrust into the public eye, have no expectation of privacy.[13] Against this logic of "consent once, circulate forever," we need to create spaces for mass loitering, in which we can be in public—we can claim public space—and be not attacked.[14]

The Friend of My Friend Is My Enemy (and Thus My Friend)

On the night of August 11, 2012, a sixteen-year-old girl was dragged almost naked from party to party in Steubenville, Ohio. Unconscious, she was fondled, fingered, and urinated on by at least two boys. A larger group, dubbed the "rape crew," took pictures and videos of "the dead body," which they then posted to *Twitter.com*, *Facebook.com*, *Youtube.com*, and *Instagram.com*. In one particularly disturbing twelve-minute video, a young man cracked jokes about the ongoing actions (including sodomy), claiming: "they raped her quicker than Mike Tyson raped that one girl."[1] The boys involved were members of the Steubenville football team: young men allegedly treated as heroes in that rusting Ohio town.

The girl, who lived in a neighboring city, and her family discovered what happened that night through images posted of her on social media. In the days that followed, the "crew" and others deleted their earlier posts, while also using social media to craft alibis.[2] Alexandra Goddard, who had lived in Steubenville and therefore doubted that anyone would be convicted, posted all the materials she could find to her blog, which eventually led to a lawsuit filed against her.[3] Partly in response to this suit, Anonymous—a free-floating group that first made its name through trolling, rickrolling, and harassing a fourteen-year-old female bully—resuscitated the deleted files and organized protests.

By all accounts, Steubenville defies rationality, but
unfortunately not due to the nature of the crime, for rape
is all too common in the United States. Rape is also
difficult to prosecute because conviction usually and
unfairly relies on the perceived virtue of the victim: the
victim must be impeccable and must be able to describe
traumatic events coherently—a nearly impossible task,
given the effects of Post Traumatic Stress Disorder
(PTSD). Steubenville became known nationally because of
the "crew's" use of social media and because of
Anonymous's involvement. Tellingly, the two young men who
were convicted received minimum sentences of one and two
years in juvenile detention. Poppy Harlow, the CNN anchor
who covered the trial of these young men, infamously
sympathized with them, saying: "These two young men who
had such promising futures—star football players, very
good students—literally watched as they believed their
life fell apart. … What's the lasting effect, though, on
two young men being found guilty [in] juvenile court of
rape essentially?"[4] Equally telling, the Anonymous hacker
who leaked the files may serve a significantly longer
sentence: up to ten years for his involvement.[5]

Steubenville raises several questions: Why are folks
increasingly documenting and publicizing their crimes? And
how did Anonymous (assuming, of course, that Anonymous is
a historically consistent entity) become a moral force
demanding state incarceration and involvement? Not
surprisingly, "Steubenville" has become shorthand for the
dangers and promises of social media. The exact meaning
of "Steubenville," however, varies with one's point of
view: either social media revealed crimes and injustices
usually covered up in small-town America, or they fostered
a dangerous vigilantism; either they revealed the
prevalence of rape in U.S. culture, or they encouraged
depraved and stupid behavior. Rather than champion any one
side of this debate, the next two chapters argue that new
media are important precisely because they call into
question the clean separation of publicity from privacy,
transparency from morality, community from hate. New media

shatter windows of privacy and security, transforming them
into high-speed optical cables that connect: again, **new
media are, at their best, wonderfully creepy.**

This wonderful creepiness of new media, however, is
perverted by attempts to contain it, to turn the Internet
into a series of poorly gated communities, in which we
think safety = corporate security. According to this
logic, privacy—traditionally the state of being deprived
of public power—becomes secrecy, the state of being
concealed, kept from knowledge. According to this logic,
security stems from encryption and transparency: encrypted
passwords, transparent identities. Social media, by making
identities indexical, supposedly render the Internet—or
at least the poorly gated communities that are social
media—a safe space. Randi Zuckerberg, marketing director
of Facebook, for instance, argued in 2011 that "people
behave a lot better when they have their real names
down."[6] Steubenville, and its many successors and
predecessors, undermine this assertion.

This trust that transparency will induce good behavior
stems from the Enlightenment and eighteenth-century
notions of discipline and publicity, which, according to
Michel Foucault, the Panopticon encapsulated. As Foucault
explains, this mythic architectural structure, devised by
Jeremy Bentham, was meant to ensure the automatic
functioning of power. The Panopticon consisted of a
central tower, which was to shelter the owner of the
Panopticon and his family, and an annular structure, which
was to house inmates in solitary confinement. Windows were
to be placed on either side of inmates' cells, so the
inmate was theoretically always in view; the central tower
was to be equipped with blinds, so the inmate could not
be sure when she or he was being watched. At first,
justice was to be swift, and inmates were to be punished
immediately for any wrongdoing. The point, however, was
to produce a state of constant vigilance in the inmates
so that they soon became their own watcher, "the principle
of [their] own subjection."[7] Incapable of verifying the

guard's gaze, inmates were to "internalize" the light and thus become good for goodness' sake. The Panopticon was designed to induce in the inmates—and the inspectors—"a state of conscious and permanent visibility that assures the automatic functioning of power."[8] Thus, as Foucault nicely summarizes, "a real subjection is born mechanically from a fictitious relation."[9] This real subjection, Foucault goes on to argue, was to make power productive: to turn power from an outside force wielded by a Sovereign subject through spectacular and uneven punishments to a general and generative force that produced productive men. The finer resolution of power offered by the disciplines was to "reduce everything that may counter the advantages of number. … [It] fixes; it arrests or regulates movements."[10] In contrast to the immobile subjects, the disciplines "swarm[ed]" and infiltrated other forms of power, making them lighter and more efficient.[11]

Although Foucault saw the Panopticon as the icon of disciplinary society, the highly mobile Santa Claus—who emerged during the same time period as the Panopticon—is arguably a better model. The children's song about Santa Claus echoes the spirit of the central tower:

> Oh, you better watch out,
> You better not cry,
> You better not pout,
> I'm telling you why:
> Santa Claus is coming to town.

> He sees you when you're sleeping,
> He knows when you're awake,
> He knows if you've been bad or good,
> So be good for goodness sake.

Santa Claus, a role and a costume that anyone can don, ensures the automatic functioning of power: it makes children "good for goodness sake." Santa's gifts are the reward for good behavior (and the denial of gifts the punishment), but the main point is to outgrow him, to become so habituated to being good that neither reward nor punishment is needed. Again: a real subjection is born from a fictitious relation. (Habit as discipline.)

Santa Claus, now a sad old man (possibly a pedophile who lurks in malls, or gets very drunk with many other Santas annually on the streets of San Francisco), has been superseded by Anonymous and the Guy Fawkes mask (a mask that anyone can wear) and its legacies. Anonymous, not Santa Claus, knows if you've been bad or good. Consider their song:

> We know what you did
> We know who you are
> We are your neighbors and your friends
> We are legion
> We do not forget
> We do not forgive
> Expect us.

The rise of Anonymous—as a dark and necessary figure for hope (as Gabriella Coleman has argued so presciently)—is linked directly to the failure of the disciplines and to rewards in an era in which everyone receives a gold star or a trophy.[12] It is also linked the failures of democracy. As Danielle Allen has put it, anonymous speech is "a solution when the costs of public speech have been set too high."[13] Anonymous—with its multiplicity, its stand against celebrity, and its embrace of "dirty" forms of dissemination, from Youtube to 4Chan—is an intriguing contrast to Snowden, whose decision to 'come out' is narrated in the Academy Award-winning documentary *Citizenfour*, in which he asks the target to "be painted on his back."[14] Both, though, are tied to transformations in the values of privacy and publicity: **privacy is being secreted everywhere**.

Arguably what is most surprising and important about Steubenville and the general publicizing of transgressions are the ways in which privacy does and does not remain. As many, including Jürgen Habermas, have written, the emergence of privacy as something to be valued is linked

to the rise of the bourgeois public sphere.[15] The rise of
    privacy as a right within the United States, as Eden
  Osucha has revealed, is linked to gender and race, for
privacy was formed as a right to protect deserving white
 women—and thus the rest of the citizenry—from the glare
of publicity: from the mass circulation of their images,
    which was described as inherently pornographic and
    racializing.[16] It is not surprising, then, that the
    Steubenvilles are 'countered' by incidents of "slut
  shaming," in which mainly white teenage girls or young
 women are publicly shamed for being 'bad' users, for no
 longer holding privacy sacrosanct, as though they, rather
than leaky networks—or more precisely, attempts to contain
  these networks—were to blame for information secretion
                            (more on this in chapter 4).

This impoverished privacy is a habit, a possibly dangerous
habit that covers over the promiscuous nature of networks,
   their wonderful creepiness. What can and should privacy
    do in an era in which the most important binary is not
  between public and private, but rather between open and
 closed, between shopping malls and gated communities? How
            can we build a genuine 'public safety?'
                                  =END Steubenville

3 The Leakiness of Friends, or Think Different Like Me

The constitutive public sphere sweeps aside as merely private all obstacles, privileges, special rights, atavisms, and peculiarities that stand in the way of the public establishment of this order.

—Oskar Negt and Alexander Kluge[1]

One of the key dimensions of the fantasy of intimate love is its stated opposition to all other forms of social determination even as it claims to produce a new form of social glue. The intimate event holds together what economic and political self-sovereignty threaten to pull apart.

—Elizabeth Povenelli[2]

Love is blind; friendship closes its eyes.

—Anonymous, or Friedrich Nietzsche

How are networks inhabited? What sustains connections, and what is the relationship between the experience and image of networks?

To respond to these questions, this chapter begins by contrasting the Internet of the 1990s, in which freedom and empowerment allegedly stemmed from an anonymity that was no anonymity, to that of the early twenty-first century, in which authentication and authenticity supposedly save users from dangerous strangers. This safety, based on the transformation of users into reciprocal and reciprocating 'friends,' was and is no safety, for online **friends are an extremely leaky technology**. Online friendship—a concept that muddies the neat boundary between public and private, work and leisure—encapsulates the promise and threat of networks: the promise of an intimacy that, however banal, transcends physical location and enables self-made bonds to ease the loneliness of neoliberalism; the threat of a security based on poorly gated 'neighborhoods.' That is, to update Margaret Thatcher's famous quip, "there is no such thing as society. There is [a] living tapestry of men and women and people," now there is a monstrous, undead chimera of 'friends' constructed through neighbor-hoods of likeness and difference.[3] This authenticating friendship, this chapter also emphasizes, perverts traditional concepts of friendship by transforming it from an essentially broadcast (and private) action—an unre-ciprocated act to love/like someone—to a banal, reciprocal, and 'authentic-like' relation.[4] Through this impoverished friendship, relations are mapped and extrapolated: habitual actions—liking, retweeting, posting, etc.—used to create profiles to carefully track, preempt and craft consumption. Regard-less of your own individual actions, **YOU are constantly betrayed by people who 'like YOU' and who are algorithmically determined to be 'like YOU.'** If cyberporn catalyzed the emergence of markets on the Internet, 'friending' practices have made the Internet itself a market of YOUs.

The Internet, as It Once (Never) Was

The popular imaginary of the Internet changed radically from the mid 1990s to the mid 2000s. This change is as extreme as the Internet's earlier transformation from a noncommercial, military, academic, and govern-mental 'public good' to a mass medium, when its backbone was sold to private corporations in the early to mid 1990s.

As outlined in chapter 2, in the mid 1990s, the Internet emerged as "cyberspace," a space of freedom and anonymity. In 1996, John Perry Bar-low infamously declared the independence of cyberspace, writing, "We [in cyberspace] are creating a world that all may enter without privilege or prejudice accorded by race, economic power, military force, or station of birth. We are creating a world where anyone, anywhere may express his or her beliefs, no matter how singular, without fear of being coerced into silence or conformity."[5] Science fiction grounded this belief in the Internet of the mid 1990s as cyberspace, as an ideal space free from physical coer-cion. Sitting at a typewriter and inspired by the world of video arcades and punk rock, William Gibson coined the term "cyberspace" in 1982, eleven years before the National Center for Supercomputing Applications (NCSA) introduced Mosaic, the first graphics-based web browser. In his 1984 novel *Neuromancer*, which fleshed out a dystopian post-World War III future in which the United States had disappeared and the world was dominated by Japanese *zaibatsus*, Gibson described cyberspace as "a consensual hallucina-tion": "a graphical representation of data abstracted from the banks of every computer in the human system."[6] In this space, elite marauding con-sole cowboys disdainfully referred to their bodies as meat.[7]

Despite the vast differences between William Gibson's vision of cyber-space and the Internet as it actually existed in the mid 1990s, this confla-tion of the Internet and cyberspace spread far beyond fans of cyberpunk fiction.[8] The notion that the Internet was fundamentally ungovernable because it was a global ethereal space grounded arguments against the enforceability of the Communications Decency Act (CDA). The famous *New Yorker* cartoon perhaps best encapsulated the dreams surrounding the early Internet as cyberspace: "On the Internet, nobody knows you're a dog" (figure 3.1). Freedom supposedly stemmed from anonymity, from no one knowing who you were. In this space, the authentic self could finally be revealed and an authentic public sphere could emerge because discrimination—which 'naturally' stemmed from the presence of raced bodies rather than racist institutions—could be eliminated. Although this idea seems incredibly naïve to us now and seemed so even then (it was

"On the Internet, nobody knows you're a dog."

Figure 3.1
New Yorker cartoon. Reprinted with permission. Peter Steiner/The New Yorker Collection/The Cartoon Bank.

criticized heavily in the mid 1990s as hype), it remains powerful: Snowden in the documentary *Citizenfour* (2014) for example, explained that he decided to go forward with his leaks because he remembered this Internet, in which children were taken as seriously as experts in online spaces. This utopian version of cyberspace promised technological solutions to political problems. Through its popularization, cyberspace bizarrely moved from foretelling a dark future to signaling a happy one, and the liberal position on technology moved from one of protest (for example, the 1980s anti-nuclear movement that informed Gibson's cyberpunk vision) to one of enthusiastic embrace.

The promise of the Internet as a bodiless public was made most clear in rhetoric surrounding the Internet as a raceless space.[9] It seemed impossible to advertise the Internet in the mid to late 1990s without featuring happy people of color singing its praises. One particularly compelling and influential commercial, MCI's *Anthem*, contained dialogue such as "There is no race," "Utopia? No, the Internet." The message of these commercials was not even the banal "Do not discriminate," but "Get online if you want to avoid being discriminated against." In addition to assuming that these raced bodies (rather than racist bodies and actions) caused discrimination, these commercials presumed a racist attitude, for they assumed that viewers would see these bodies and then automatically understand why these 'others' would be happy to be on the Internet. These representations also did not reflect then-current Internet demographics, the same corporations that touted the Internet as the great equalizer also sponsored round tables on the digital divide. The point of this rhetoric, however, was not to get more people of color online, but rather to get the 'general public' online. These dreams were strongest and this imagining most compelling at a time when very few people were on the Internet. Indeed, once there existed a certain density of Internet users, these commercials disappeared (they lasted longer in Japan, where the initial Internet uptake was slower than in the United States). Cisco Systems' *Are You Ready?* television advertisements made this logic clear: they featured interchangeable people from around the world, who accosted viewers with predictions about worldwide Internet usage and asked, "Are you ready? We are." The phrase "We are" revealed the threat behind these seeming celebrations: get online because these people already are. It was "we" against "you."

As I argued in *Control and Freedom*, this notion of the Internet as a medium of the mind, in which body and soul, physical and mental location, could be separated, relied on a very odd understanding of the Internet—one that ignores the actual operations of TCP/IP (the control protocol that is the Internet). As any packet sniffer quickly reveals, our computers are engaged in constant, incessant, and promiscuous exchanges of information, without which there could be no communications at all. That book thus started with the question: Given that the Internet is one of the most compromised and compromising forms of communication, why has it been bought and sold as empowering and freeing—as a personalized medium?

The question that drives this book and this chapter is different, for the dominant imagining of the Internet in the early twenty-first century has moved away from this odd understanding of the Internet, with its intimately intertwined dreams of cyberspace and virtual reality, toward another equally strange one. In the first decade of this century, with the advent of Web 2.0, the Internet has become a semipublic/private space of 'true names' and 'authentic images' (figure 3.2). Rather than being a form of virtual reality, the Internet is now viewed as augmenting reality. In this semiprivate or semipublic space, freedom stems not from anonymity, but rather from knowing who is a dog and who is not.[10] The authentic increasingly stems from the privately authenticating, from what I term YOUs value.

Figure 3.2
Image of author posted by friends to Facebook

Crucially, this new version of the Internet envisioned trusted social interactions as based on transparency and conceived of the default user not as a lurker but a friend. This move toward transparency responded to the failures of the initial Internet to live up to its hype as an ideal marketplace of ideas, as the Athenian agora come true. By the early 2000s, the early promises of the web were exposed for what they were: unfulfilled and perhaps unfulfillable imaginings. Like the newsgroups that preceded them, chat rooms were often nasty spaces subject to "Godwin's Law," and open listservs were dying, killed by spam and by trolls, whose presence was amplified by those who naively 'fed' them and others who admonished them for doing so.[11] Further, the Internet was filled with phishing scams, and seemingly private email accounts were flooded with spam messages advertising pornography, body modification tools, and dodgy pharmaceutical companies.

In particular and very early on, child pornography was seized upon as encapsulating the threat of the Internet (it has now been supplemented by terrorism). As discussed earlier, the first attempt to regulate the Internet's content coincided with its deregulation: the Communications Decency Act (CDA) of 1995, an act that passed the U.S. Senate with an overwhelming majority after senators perused tightly bound printouts of "perverse" images that Senator James Exon's 'friend' had downloaded for him.[12]

Pornography was, and still is, central to the two issues that map the uneasy boundary between public and private: regulation and commerce. The Internet's privatization paved the way for "cyberporn," to the extent that it made digital pornography a hypervisible threat/phenomenon. Cyberporn, in turn, paved the way for the "Information Superhighway" to the extent that it initiated the Internet gold rush and caused media, governments, and commercial companies to debate seriously and publicly the status of the Internet as a mass medium. Before the Internet went public through its privatization, legislators had shown no concern for minors who accessed the alt.sex hierarchy, or who logged onto "adult" bulletin board systems (BBSs); pornography's online presence was so well known among users it did not even qualify as an open secret. Upon 'discovering' the obvious, the media and politicians launched a debate about 'free' speech focused on assessing, defining, and cataloguing pornography. The impact of this debate on the Internet was profound. Although the CDA eventually was ruled unconstitutional, debate over it and its credit card–based safe haven provisions helped foster the first successful online businesses: commercial pornography sites. Actors on both sides of the CDA debate portrayed these sites as responsible, rather than greedy, for charging for information that had been freely accessible for years (even though minors could easily get access to a credit card). The profitability of these sites was key to convincing less risqué businesses that, contrary to initial public wisdom and skepticism, an online marketplace was possible: news media, such as CNN, reported that secret consultations occurred between porn web mistresses and representatives of mainstream businesses, such as IBM, seeking to create an online presence.[13]

The CDA was eventually defeated, in part due to an overriding belief in the Internet as a medium of freedom and anonymity. According to Judge Stewart Dalzell, who granted a temporary injunction against the CDA, the Internet proved true Justice Oliver Wendell Holmes's famous assertion that "the best test of truth is the power of the thought to get itself accepted in the competition of the market."[14] More soberly, Justice John Paul Stevens concluded his decision upholding this injunction by celebrating the phenomenal growth of the Internet and declared, "the interest in encouraging freedom of expression in a democratic society outweighs any theoretical but unproven benefit of censorship."[15]

Many, although not all, now blame anonymity, which once grounded the dreams of the Internet as a utopian space of the mind, for destroying the possibility of a civilized public sphere. Corporations such as Google and Facebook, which needed and still need reliable, authenticated information for their data-mining operations, promoted the tethering of online and offline identities as the best way to foster responsibility and combat online aggression.[16] Randi Zuckerberg, marketing director of Facebook, argued in 2011 that, for the sake of safety, "Anonymity on the Internet has to go away." Eric Schmidt, CEO of Google, made a similar argument in 2010 stating, "in a world of asynchronous threats, it is too dangerous for there not to be some way to identify you."[17] These arguments were not new or specific to Web 2.0: ever since the Internet emerged as a mass medium in the mid 1990s, corporations have framed securing users' identities as crucial to securing trust. Two assumptions drive this argument: one, that the worst dangers come from anonymous strangers rather than friends; and two, that transparency guarantees better actions (again, Zuckerberg's opinion that people act better if they give their real names).

Many scholars have challenged this linking of trust and security, most insightfully Helen Nissenbaum. Writing in 2001, Nissenbaum noted that, although security is central to activities such as e-commerce and banking, it "no more achieve[s] trust and trustworthiness, online—in their full-blown senses—than prison bars, surveillance cameras, airport X-ray conveyor belts, body frisks, and padlocks, could achieve offline. This is so because the very ends envisioned by the proponents of security and e-commerce are contrary to core meanings and mechanisms of trust."[18] Trust, she insists, is a far richer concept that entails a willingness to be vulnerable. The reduction of trust to security assumes that danger stems from outsiders, Nissenbaum writes, rather than from "sanctioned, established, powerful individuals and organizations."[19]

The development of the Internet has made Nissenbaum's words prophetic. With 'transparency,' we have seen not only an explosion of e-commerce but also a blossoming of dataveillance, cyberbullying, and cyberporn. The naïve presumption that transparency would 'cure' the evils of the early Internet—pornography, trolling, flame wars, etc.—has proven to be false. Cyberbullying takes place most effectively within the trusted structure of 'friend' networks, for it is most traumatic when both parties are known or are assumed to be 'friends of friends.' Steubenville, and many more events like it, involved real names.[20] Further, child pornography is expanding—figure 3.3 has come true—but not in the terms initially imagined. Rather than being produced by lecherous old men for lecherous old men, it is being produced by teenagers for teenagers: participatory culture in flagrante delicto.[21] Child pornography, through sexting, has become crowd-sourced. The fact that teenagers are the produsers (producers + users) of these images, though, does not protect them from prosecution. In one early (2004) high-profile case in Florida, A.H., a sixteen-year-old girl, took photographs of herself and her seventeen-year-old boyfriend J.G.W. having sex. The sex was consensual and legal, and these photographs were never distributed to a third party. Regardless, both minors were convicted of knowingly producing, distributing, and promoting child pornography.[22] In Florida, they could marry; they could have sex; but, due to laws introduced in the 1990s to protect them from the Internet, they could not legally take images of themselves having sex without becoming sex offenders. So, even though their names cannot be published because they were minors at the time, they will have to register as sex offenders for the rest of their lives. Intimate expression is now imbricated in public structures in ways that unsettle our long-standing conceptions of privacy and security.

Most forcefully, this reveals that the desire for a reciprocal and authenticating, if not entirely authentic, type of intimacy—for friendship as 'friending'—makes the Internet a more deadly and nasty space, in which we are most in danger when we think we are most safe and in which we place others at risk through the sometimes genuine care we show for them. Given this, we need to ask ourselves: What exactly does 'friending' do? How have we all become friends?

Figure 3.3
July 3, 1995, cover of *Time Magazine.* Reprinted with permission.

The Friend of My Friend Is My Enemy

How and why did we suddenly become friends online? What kind of notion of friendship drives online networks?

In the early years of this century, *Friendster.com* made popular the notion of an online friend within the United States. Users of the site created a profile page that consisted of spaces for testimonials and for a listing of one's friends. Conceived of as a dating site, *Friendster.com* was launched in beta mode in 2002 and was initially popular with three subgroups: attendees of Burning Man, gay men, and bloggers, all mainly living in San Francisco and New York. As danah boyd has revealed, the site quickly spread to other subgroups such as goths, ravers, and hipsters, and it garnered mainstream media attention by mid 2003. By October 2003, *Friendster.com* had more than 3.3 million accounts.[23]

The concept driving *Friendster.com* was simple: to compete with sites such as *Match.com*, it relied on semipublic declarations and testimonials by friends, rather than the results of extensive and complex surveys. Further, it leveraged already existing connections to create a user base larger than those actively seeking romance; the resulting instantiation of matchmaking was presumably more effective than its offline variant because it revealed more connections between friends than possible through purposeful matchmaking. These 'friends' were not only an important source of connections; they were also potential users of the match-making aspects of the site. The site exploited the muddy boundary between those looking/open and those taken/closed, while also seeming to respect this boundary by asking people to state their relationship status. Users were not offered a universal view of the site, nor access to all profiles, but were forced to navigate an egocentric network limited to profiles within four degrees of separation (friends of friends of friends of friends). This four-degrees limit was inspired by the sociologist Stanley Milgram's classic experiment, in which he (allegedly) showed that most people are connected within six degrees of separation. To maintain its legitimacy as a dating site, *Friendster.com* thus thrived and depended on authenticity and authentication of one's identity and character by one's friends.

The site, however, soon fell out of favor in the United States. By 2004, the majority of users were from Singapore, Malaysia, and the Philippines. boyd, among others, has linked the demise of *Friendster.com* to the "Fakester Genocide," a concerted effort by the *Friendster.com* management to delete the accounts of fakesters: people who created fictional accounts of things (such as Burning Man) or people (such as Angelina Jolie).[24] The management deleted these popular accounts because they viewed these fakesters as undermining the theoretical premise that grounded the site. According to founder Jonathan Abrams, "fake profiles defeat the whole point of Friendster. ... The whole point of Friendster is to see how you're connected to people through your friends."[25] By linking to a popular fakester—by joining a community of fans—a user quickly became connected to many people who were not connected to him or her via a 'real' friend. Thus, according to Abrams's logic, because these fakesters were so promiscuous, they thwarted the correlation between *Friendster.com* and real life. (This assumption ignored the fact that mutual interest in a fakester could serve the same purpose as being a friend of a friend of a friend of a friend (or a user's neighbor); it also revealed that authentication was valued over commonality.)

Besides undermining the theory driving the site, the fakester phenomenon demonstrated the potential for community to become a network weapon: it seriously challenged *Friendster.com*'s technology because, with so many connections, the site ground to a halt. Deleting fakester accounts, though, led to an exodus not only of those violating the site's conditions of use, but also those sympathetic to the fakesters and those disconcerted by the heavy-handed tactics of the *Friendster.com* management. Still others left because others had because, without constant activity on testimonial boards, the site became boring and profiles became "frozen" relics of past conversations.[26] The mass exodus revealed the *Friendster.com* management's blindness to a basic social media fact that their site had encapsulated: online, to be is to be updated. Constant updates by others and oneself maintain online presence. As *Facebook.com*'s newsfeed and livestream have made clear, sharing surveillance with users not only makes users more comfortable with surveillance, it also makes them engage more with the site. The constantly changing newsfeed keeps the site 'alive'; making users' actions public keeps SNSs (social networking sites) from appearing frozen.

Despite its demise as a dating site, *Friendster.com*'s legacy is its popularization of a bizarre notion of friendship as reciprocal and verifiable—that is, a matter of mutual agreement. This notion of friendship resonates strongly with the concept of 'love' in neoliberal societies, as they both focus on and amplify the importance of individual decisions. As anthropologist Elizabeth Povinelli has asserted, "One of the key dimensions of the fantasy of intimate love is its stated opposition to all other forms of social determination even as it claims to produce a new form of social glue. The intimate event holds together what economic and political self-sovereignty threaten to pull apart, and it does so while providing an ethical foundation to a specific form of sex; stitching the rhythms of politics and the market to the rhythms of the intimate subject."[27] Intimate love, that is, builds new bonds that integrate individuals—released by love from other bonds, such as family—into the logic of the market. As she further explains: "Because this kind of self-transformation leans on the openness of other people to the same type of self-transformation, autological intimacy functions as a proselytizing religion. Like capital, intimacy demands an ever-expanding market; and, like capital, intimacy expands through macro-institutional and micro-practices."[28] As *Friendster.com* revealed, the desire for love spreads quickly and creates networks, but friendship expands even more rapidly, its weakness compared to love its strength. Friends embody the neoliberal spirit of independent choice, for we choose our friends in ways we cannot choose our family or those with whom we fall in love. Through friends we craft our affiliations and our selves. Already weaker than love, friendship is made even weaker by its transformation from what Derrida and others have described as a nonreciprocal relation to a benign mutual agreement: a click that confirms that one recognizes the other.[29] It is also, however, a click—a response to a call—that, as Emmanuel Levinas has claimed, makes one hostage to the other.[30]

As boyd notes, this impoverished notion of friendship—which reduces all sorts of relationships to YES/NO friendships—creates all kinds of dilemmas. Most particularly, it compromises the separation of work from leisure, family from friend. Yet these boundary crossings, and the crises they provoke, are not merely unfortunate side effects; they are the point. As boyd herself comments, the purpose of a *Friendster.com* friend was to confuse the boundaries between public and private: it "public[ly exhibited] private relationships in order to allow for new private interactions."[31] These publicly exhibited private interactions complicate traditional understandings of the public sphere. The site's stretching of the notion of a friend was also key to its expansionist logic. To provide the best authentication and the most variety, it had to move beyond normal notions of friendship. Further, this compromising of the boundary between work and leisure was its business model. Through acts of 'friending,' writing on people's walls, etc., content is freely provided for these sites, and the connections between users, which SNSs profit from, are revealed. 'Friending' is a key part of what Tiziana Terranova has called "free labour."[32]

Social networking sites that followed in the wake of *Friendster.com*, such as *MySpace.com* and *Facebook.com*, inherited the dilemmas and benefits of this banal, and therefore even more dangerous, definition of friend. The early evolution of *Facebook.com* and its relation to its print predecessors reveal the attraction and power of enclosed open spaces, that is, spaces that feel 'free' because they are enclosed. Facebooks are traditionally print publications circulated at liberal arts institutions that feature pictures of students along with some information about them. These books were key to creating something like community, that is, transforming "friendly strangers" into friends, or at the very least acquaintances, whom one knew by name. Various eating houses at Princeton would invite freshmen to their parties based on their pictures in the *Freshman Facebook*; faculty at Harvard were given facebooks with the hope that they would learn the names of their students. Facebooks were thus key to that perhaps wonderful, perhaps creepy sense of community for which one pays so dearly at these institutions. *Facebook.com* was started by a Harvard undergraduate, Mark Zuckerberg, in 2004 to give students in various Harvard residence halls a way to identify and meet other students. It soon spread to other universities in the Boston area and to other Ivy League colleges. Within two years, there were at least 9 million users of *Facebook.com* in countries including Canada, England, India, Mexico, and Australia; by 2005, 85 percent of undergraduates in supported schools used *Facebook.com*, and 60 percent logged on at least once a day.[33] There were also corporate and high school versions of *Facebook.com* before *Facebook.com* gave up its school and corporate-based centers and began to adopt a more egocentric focus, similar to *Friendster.com* and *MySpace.com*.[34]

As the term "portal" makes clear, the early boundaries that *Facebook.com* established were as crucial as the connections they enabled, for a portal is an elaborate door that takes people inside, not outside (see figure 3.4). *Facebook.com* enabled users with valid email accounts to create personal profiles, which, in 2005, could contain photos, birthdates, phone numbers, courses taken, and descriptions of one's religious and political views. To view a detailed profile, though, you needed to be on the same network as the profile's and the profile's default had to be set to allow network access (the network could be as broad as Brown University or as narrow as faculty at Brown), or you and the profile owner had to be 'friends.' You could create a blog or leave messages on your friends' walls, as well as create a newsfeed detailing your latest actions. Like many Web 2.0 applications, *Facebook.com* actively discouraged anonymity. Although there was no way

Figure 3.4
Schottenportal, Scots Monastery, Regensburg, Germany. Photo by Richard Bartz.

of verifying student photographs, many of them were framed as self-representations, and their sometimes compromising natures reeked of authenticity. Race, through these photographs, made itself visually present; there was no attempt to create a race-blind space, but rather a desire to make the Internet reflect—or better yet refract—one's physical space.

Social networking software such as *Facebook.com* and *Friendster.com* fostered a link between the online and the offline: one used *Facebook.com* to track down a fellow student. *Facebook.com* was thus popularly imagined as a safe space, that is, a network that one could control. The free flow of information—the cell phone numbers, the frank assessment of their classes, and compromising photographs that students would post—depended on a gated community that supported existing networks of privilege. Initially, for instance, a student at Texas A&M could not see the profiles and networks of Harvard students, unless she was explicitly friended by a Harvard student. Even then, she was offered access not to the entire Harvard network but to an egocentric network that depended on the default settings of other users. Rather than being an open university, *Facebook.com* was a geographically bound, sheltered space. The site took the perceived intimacy, so crucial to liberal arts institutions, to the 'masses' (larger public institutions) by creating semiprivate enclosures within seemingly open spaces.[35] *Facebook.com* was thus initially less appealing to high school students, who tended already to know people in their classes, than *MySpace.com*, which also required one to have a profile before one could view others but offered a more expansive environment. Everything changed when *Facebook.com* opened up its network, which it could do because of the sheer density of 'friends' it had enclosed.

As mentioned earlier, for all their rhetoric of safety and enclosure, *Facebook.com* and other SNSs, which enable users to 'choose' their friends and their level of exposure, are hardly safe spaces.[36] These sites reveal the lie of 'stranger danger.' As many studies have demonstrated, teenagers, with the exception of gamers, marginalized, and "at-risk" youth, do not develop friendships with strangers online.[37] Rather, as boyd writes, they use these tools to "maintain preexisting connections, turn acquaintances into friendships, and develop connections through people they already know."[38] Yet this maintenance has not meant that young adults have been free of harassment while in these spaces. Instead, because these sites also amplify offline relations by expanding the temporal and spatial range of their users' interactions, they have expanded the range and force of bullying. The pressure to accept a friend request—in order to be polite or to access another's 'private' profile—means that one is likely to be 'friends' with an enemy (again, to the extent that friends and enemies can be distinguished).[39]

Sexting among juveniles also reveals the fact that identification and authentication can further a more virulent "peer-based" harassment. As the 2012 report *A Qualitative Study of Children, Young People and "Sexting"* explains, the often-coercive demand for sexts comes from peers rather than unknown strangers.[40] Although sexting, especially within the United States, has far more dire legal consequences than cyberbullying (as shown in the example cited above), it is more common and was initially considered less upsetting than cyberbullying.[41] Sexting usually occurs on devices and through modes of communication imagined as even more private than SNSs: text messaging on one's mobile phone. Texting is considered more secure than SNSs such as *Facebook.com*—also called "baitbook"—even though it is easy to take and circulate screen shots of mobile phones.[42] The British authors of *A Qualitative Study of Children, Young People and "Sexting"* focused on the role of BlackBerry Messaging (BBM) in the spread and circulation of sexting. Again, perceived security leads to greater danger. Through BBM, which relied on secure PINs—and thus appeared extremely private and secure—demands for sexts became unrelenting, and these images were quickly distributed beyond their intended recipient. Through authentication, teens became involved in a serious criminal activity: the production of child pornography.

Given the failures of transparency and authenticity to guarantee safety, why did and does the concept of online friendship persist? What is at stake in 'friending?' 'Friending' and its creation of new voluntary and involuntary bonds have been crucial to the transformation of the Internet into a market: into a "Big Data" goldmine through the creation of affiliation networks.

YOU Matter, or If YOU See Something, Say Something

How is value generated online? However disappointing and deferred, why were Facebook's IPOs worth so much?

At a certain level, the answer seems simple. Value is generated online, and networks are valuable because information has become a commodity. As many scholars, including Manuel Castells, have argued, information has moved from an entity necessary for production to a product in and of itself.[43] Across the political spectrum—from Marxists to neoliberal capitalists—information/knowledge is portrayed as a valuable immaterial commodity. To those who seek to expand and exploit capitalist markets, information is power: It is valuable because with it, users can make the right (that is, profitable) decisions.[44] Goldman Sachs is therefore willing to pay millions of dollars to develop software that can process information and help it to make decisions microseconds before its competitors. On the left, scholars such as Tiziana Terranova and Alexander Galloway have emphasized the link between immaterial commodities and labor processes. Terranova, focusing on the ongoing labor central to the success of any website, argues that with information, "the commodity does not disappear as such; it rather becomes increasingly ephemeral, its duration becomes compressed, it becomes more of a process than a finished product. The role of continuous, creative, innovative labor as the ground of market value is crucial to the digital economy."[45] Both on the right and the left, then, there is a sense that the value of information depends on timing.

The value of information, though, is not simply tied to its newness or initial discovery, and in this sense, the timing of networked information differs from that of its print predecessors, such as newspapers. Whereas Walter Benjamin, comparing the times of the story and of the news, could once declare, "the value of information does not survive the moment in which it was new. It lives only at that moment; it has to surrender to it completely and explain itself to it without losing any time," now newness alone does not determine value.[46] In the early twenty-first century, news organizations began charging for old information. In 2015, the *New York Times* online, for example, offered a certain number of current articles for free, but charged for its archive; similarly, popular mass media shows such as *This American Life* provided only the current week's podcast for free. Users pay for old information either because they want to see it again or because they missed it the first time, their missing registered by the many references to it (consider, in this light, all the *Youtube.com* videos referencing *Two Girls, One Cup* after that video was removed). Repetition produces value; repeated references and likes by friends and strangers mark something as valuable, as worth visiting, as worth downloading. Information— some event, incident, media object, etc.—becomes valuable when it moves from a singularly noted event to one that elicits 'mass' response (when it becomes 'viral'). This is why sociological analyses of sites such as *Twitter. com* take as their base unit retweets, likes, and other repetitive acts.

As this repetition makes clear, value is not generated by one YOU but rather by a plethora of YOUs: by the very interconnections between the various YOUs. YOU, again, is central to the operation of networks because it is both singular and plural. In its plural form, it still refers to individuals as individuals, rather than creating another communal subject, a 'we,' from more than one 'me.' In a network, that is, the nodes are still theoretically distinct, however aggregated. This YOUs value is related to and differs from other notions of networked value, which emphasize the importance of crowd sourcing, peer-to-peer production, and the collaborative nature of knowledge, concepts that have been developed insightfully by scholars such as Yochai Benkler, Pierre Lévy, and Paolo Virno.[47] Whereas these notions emphasize the collective effects of voluntary actions, YOUs value emerges through the mainly involuntary effects of voluntary and involuntary actions, from searches to mouse clicks, from likes to posts. It is also produced by a certain politics of storage that makes possible affiliation networks, which trace and link users' online actions. If our world is data rich, it is not simply because users provide content for free, but also because every interaction is made to leave a trace, which is then tied to other traces and used to understand YOU, where YOU is always singular and plural. Whether any particular YOU is aware of it or not, YOUs constitute a latent resource. *Facebook.com*, *Amazon.com*, and *Google.com*, among other sites, mine user data not simply to identify unique users but also, and most importantly, to see how their likes, etc., coincide with those of others. Collaborative filtering algorithms developed by *Netflix.com* and *Amazon.com* to recommend purchases and classify users exemplify this, for they analyze and collect data in ways that suspend the difference between the individual and collective statistical body, even as they respect and insist on this difference by providing users with individual logins and pages optimized for them. This is why SNSs seek to be portals, for enclosing users within spaces is the easiest way to analyze and track these connections. Initially, *Amazon.com* filtered according to content: it made suggestions for further purchases based on what YOU and others have bought; *Youtube.com* made its recommendation based on covisitation counts—that is, how many users watch any two videos back to back.

Early on, *Netflix.com* relied on collaborative filtering: it filtered according to similarities between films and viewers. This was no easy task because, as Mung Chiang explains in *Networked Life*, the data is both very big and sparse—there are millions of subscribers and films, and yet very few users rate films.[48] To improve its recommendation system, *Netflix.com* famously issued a challenge: it offered a large chunk of its database and a lot of money to whoever could improve its recommendation system by 10 percent.[49] The winning algorithm employed the average rating and factors to compensate for user and movie bias; and, most importantly, it created "neighborhoods" based on the relationship between films and users. Intriguingly, a "neighborhood predictor" factored in both strong likes and strong dislikes—what mattered was how much a user deviated most from the norm with others. This use of the term "neighborhood" is telling, as it reveals once more the transformation of the Internet into a series of gated communities. The segregation of films and users into neighborhoods based on strong likes and dislikes assumes that neighborhoods are forms of voluntary segregation— that YOU reside with people 'like YOU,' whose actions preempt and shape YOUR own. This is redlining on an entirely different level; as I pointed out in chapter 1, network analytics engage in discrimination under the cover of seemingly neutral proxies that target intersections of race, class, gender, and sexuality. These algorithms make no attempt at desegregation, at expanding one's point of view by exposing one to things that are radically different. Rather, YOU reveal YOUs, where these YOUs are closely lumped together, and YOUs are defined—whether or not users speak—through YOUR affiliations.

This intersection of data and methods designed to identify individuals and those to identify larger trends suspends the traditional separation between the two archival logics to incorporate the body that Allan Sekula influentially theorized in relation to the production of photographic evidence.[50] The first, derived from the work of criminologist Alphonse Bertillon, focused on identifying the individual, on inscribing the body in the archive (figure 3.5). The other archival logic, derived from the work of the eugenicist Sir Francis Galton, sought to identify the hidden type driving the body and thus to embed the archive in the photograph (figure 3.6). Currently, these processes have become inseparable at the level of data capture and storage. The same process captures the data necessary to identify individuals as singular and to identify their relation to certain groups. *Amazon.com*, for instance, tracks individual purchases not only to create a record of a user (a digital fingerprint), but also so that it can connect that user's actions with those of others in order to make suggestions for further purchases—that is, so it can predict and encourage future behavior that conforms to, and confirms and optimizes, statistical network analyses.

These algorithms and this mining assume that the data being gathered is reliable; that users' online actions are as indexical as their body measurements and mug shots. Tellingly, *Netflix.com* did not employ the winning algorithm in all its complexity, but rather turned to Principal Component Analysis (PCA) because its database became much richer once it began to stream films.[51] To help ensure this correlation, which values users' actions over their words or ratings, websites create login structures that link a person to an ID. They also benefit from the ways in which users' friends—their likes, their posts, their tags, their retweets (or via Gmail, their email messages to us)—authenticate a user and enmesh the user more thoroughly into these networks. Their actions also help target messages directed blindly toward users and 'register' their accuracy, even if there is no direct response.

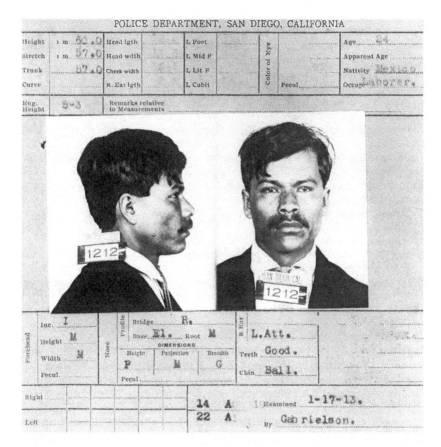

Figure 3.5
Bertillon card, 1913, reproduced in Allan Sekula's "The Body and the Archive," *October* 39 (Winter 1986): 3–64.

Whether or not YOU are aware of it, YOU are always following the mantra: If YOU see something, say something.

This mode of targeting, of the production of YOUs, resonates strongly with Louis Althusser's theorization of ideology. Drawing from the work of Jacques Lacan, Althusser argued that ideology "represents the imaginary relationship of individuals to their real conditions of existence."[52] By this, he did not mean that ideology is simply a figment of one's imagination, for one usually can only access one's real conditions of existence via the imaginary.[53] Further, the ego emerges through a process of imaginary identification with others (and other images). This identification is also always a

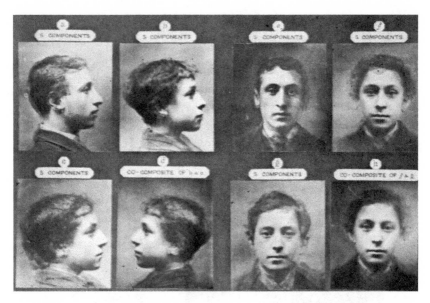

Figure 3.6
Sir Francis Galton's composite of "The Jewish Type," 1883, reproduced from Karl
Pearson, "The Life, Letters and Labours of Francis Galton," plate XXXV, in Allan
Sekula's "The Body and the Archive," *October* 39 (Winter 1986): 3–64.

misidentification, for one is not actually those other people or images (a
mirror reflection both is and is not the body). Ideology, Althusser stresses,
interpellates individuals as subjects. To explain this, he describes a police-
man yelling "Hey you!" at a person on the street. He continues, "Assuming
that the theoretical scene I have imagined takes place in the street, the
hailed individual will turn round. By this mere one-hundred-and-eighty-
degree physical conversion, he becomes a *subject*. Why? Because he has
recognized that the hail was 'really' addressed to him, and that 'it was *really
him* who was hailed' (and not someone else)."[54] By recognizing oneself as
hailed, one makes the YOU and the I coincide and thus emerges as a subject
of and to the law, and of and to society more generally. Althusser stresses
that this hailing rarely misses its mark: "Experience shows that the practical
telecommunication of hailings is such that they hardly ever miss their man:
verbal call or whistle, the one hailed always recognizes that it is really him
who is being hailed."[55] Ideology, as a telecommunicational event, provokes
a response, which is also a recognition.

Richard Dienst has revised Althusser's notion of ideology to engage more
closely with telecommunications.[56] Stressing the "hardly ever," Dienst

argues that if ideology can miss its mark, then it must "have been unstable as sign and as event, it can never simply be the transmission of a meaning to the subject." He goes on to contend that "ideology neither hails nor nails the subject in place, since both terms [subject and individual] only appear on the occasion of a telecommunication."[57] To think through this concept of ideology as an event or telecommunication, Dienst draws from Derrida's work on misdirection in *The Post Card*. Bringing together Derrida's and Althusser's arguments reveals that "ideology must be conceived as a mass of sendings or a flow of representations whose force consists precisely in the fact that they are not perfectly destined, just as they are not centrally disseminated. Far from always connecting, ideology *never does*: subjects look in on messages as if eavesdropping, as if peeking at someone else's mail."[58] There is always a distance between the "I" and the "Hey you." This does not mean, however, that ideology does not work; rather, "ideology requires a short circuit between the singular and the general so the *reception of a representation* becomes a sending back—a *representation of a reception*" (italics in original).[59] This sending back, which closes the circuit, also short-circuits the singular with the general, so that individuals respond to a general call as if it were directed at them in particular: the flows of messages create what are later identified as agents. Importantly, these flows are not directly addressed to these agents, but by responding to these calls (as when users seek out the message that they missed such as, for instance, the *Youtube.com* videos referencing *Two Girls, One Cup*), audiences become imbricated in particular expressive communities and systems of meaning.

Although Dienst developed his theory in relation to broadcast television, this description of ideology as short-circuiting the singular and the general describes the ways in which YOUs value works. In networks that track users, their captured actions involuntarily send back representations of receptions. In this sense, there is no "silence of the masses" in new media because interactivity thrives on constant, involuntary, and traceable exchanges of information.[60] This technological closing is complemented by another: through acts of friending, following, liking, and recommending, users register these receptions and they are also registered by others. Through these gestures, messages become both more directed and less general (better targeted) because users answer their friend's eavesdropped calls because they care.

Importantly, though, we still need to think through the ways in which these acts of friendship can overwhelm and compromise attempts at enclosure and attempts to master YOUs value. First, these associational links produce too much information. In the era of "Big Data," it is impossible to address all trends and information. As a 2012 article in the *Wall Street Journal* opined, "the problem that a lot of companies face is that they don't know what they don't know."[61] Second, in addition to producing controversies within these sites, caring actions also threaten to overwhelm the network and spread viruses and spam. As I argued in the last chapter, through users' efforts to foster safety, they spread retrovirally, thus defeating their computer's usual antiviral systems.

To be clear, though, I do not simply want to condemn the desire for intimacy and its dangers, for that desire is perhaps what is wonderful and productive about the Internet as well. A fascinating corollary to *Friendster.com* was the emergence of flash mobs, which emerged at the same time (2003) in New York City. As an example, consider this email message inviting a group of mainly youngish hipsters to take part in MOB #4 in New York City:

Date: Wed, 9 Jul 2003 16:40:21-0700 (PDT)
From: The Mob Project
To: themobproject@yahoo.com
Subject: MOB #4
You are invited to take part in MOB, the project that creates an inexplicable mob of people in New York City for ten minutes or less. Please forward this to other people you know who might like to join.
FAQ
Q. Why would I want to join an inexplicable mob?
A. Tons of other people are doing it.
Q. Why did the plans to MOB #3 change?
A. The National Guardsmen with machine guns had something to do with it.
Q. What should I do with my MOB $1 bill?
A. Spend it, if you like. But you may be asked to make another, for a future MOB.
Q. Can we do a MOB downtown, for a change?
A. Sure.
INSTRUCTIONS—MOB #4
Start time: Wednesday, July 16th, 7:18 pm
Duration: 10 minutes
(1) At some point during the day on July 16th, synchronize your watch to http://www.time.gov/timezone.cgi?Eastern/d/-5/java/java. (If that site doesn`tworkforyou,tryhttp://www.time.gov/timezone.cgi?Eastern/d/-5.)
(2) By 7 PM, based on the month of your birth, please situate yourselves in the bars below. Buy a drink and act casual. NOTE: if you are attending

the MOB with friends, you may all meet in the same bar, so long as at least one of you has the correct birth month for that bar.

January, February, March: Puck Fair, 298 Lafayette St. (just south of Houston). Meet just inside the front door, to the right.

April, May, June: 288 (a.k.a. Tom & Jerry's), 288 Elizabeth St. (just north of Houston). Meet in the back to the left, by the jukebox.

July, August, September: Bleecker St. Bar, 58 Bleecker St. (at Crosby). Meet in the back to the right, by the jukebox. ...

(3) Then or soon thereafter, a MOB representative will appear in the bar and will pass around further instructions.

(4) If you arrive near the final MOB site before 7:18, stall nearby. NO ONE SHOULD ARRIVE AT THE FINAL MOB SITE UNTIL 7:17.

(5) At 7:28 you should disperse. NO ONE SHOULD REMAIN AT THE MOB SITE AFTER 7:30.

(6) Return to what you otherwise would have been doing, and await instructions for MOB #5.

The first flash mob converged on the rug department of a Macy's department store; the fourth overran a Soho shoe store. As mass acts of benign communal action, flash mobs were one's friends lists come to life: ephemeral interventions into quasi-public or at the very least open spaces, enacted by familiar strangers; latent publics, activated.[62]

Intriguingly, although the organizers constructed these mobs to be as banal as possible—engaged in actions such as shopping for shoes and placed in the "safest" of public spaces (the third New York flash mob moved from Grand Central Station to the lobby of the Hyatt Hotel because of the presence of "National Guardsmen with machine guns")—they were still treated with great suspicion. As the then-anonymous New York organizer "Bill" noted, "There seems to be something inherently political about an inexplicable mob."[63] Indeed, the gathering of a mob, speaking in a language not entirely understandable in the words and gestures of official politics (that is, in a language not easily recognized as political), recalls the traditional "noisy" claiming of rights.[64] The fact that these flash mobs were deliberately nonpolitical and couched in terms of play and yet were so disruptive—coupled with the fact that they would later mutate into highly orchestrated commercial public relations events and criminalized swarms, as well as TXTmob—also exemplifies the dangers of occupying and opening this liquid space between public and private, the dangers and possibilities also exemplified by the opening that is a friend.

As well as considering the political possibilities opened and shut down through 'friending,' we also must explore other nonreciprocal modes of relation, which do not demand that ties between agents be explicitly acknowledged or bidirectional. Further, we need to consider how involuntary acts of spamming might be key to embracing the possibilities for community and action. I thus conclude with a personal anecdote to start us in this direction.

Spam, or Another Way to Say I Love You

In 2013, I fell victim to a phishing attack. The term "fell victim" is a little strong, for as soon as I clicked on the link, I knew something was wrong, and, had I not been distracted by two small children and using my iPhone at the time, I would never have made that mistake. This attack taught me what I should have already known: there is no innocent surfing online; babysitting is dangerous.

This attack, however, was brilliant: it was one of the most successful on *Twitter.com* to date. It consisted of a "private message," poorly typed and seemingly urgent, from a follower stating, "i cant believe this but there are some real nasty things being said about you here gourl.kr/Ag9hlR." I received this message from a former student, who also ran an important collaborative website, and I had just returned from a conference: The circumstances were perfect, even though the spelling errors and language should have signaled the falsity of this message (this student was far too professional to send such a message). This phishing attack did not just compromise my *Twitter.com* account; it also led to everyone following me on *Twitter.com* to be phished in turn, so it outed me as being naïve and possibly paranoid.

Predictably, many folks contacted me directly letting me know what I already knew—that I had been phished—and I had to amplify my public embarrassment by contacting everyone else and letting them know that the "private message" I had sent them was anything but. This experience made me realize that I had been taking the wrong approach to social networking. Clearly, I should only friend and follow people I hate.

There was, however, a surprising upside to this that made me decide not to take this new approach. Given that I hardly ever tweeted, the phishing attack allowed me to reach out to people who cared enough to skim over 140-character comments I might make. Spam, or phishing, became another way to say I love you.

One particular exchange made this point to me. A close colleague of mine received my phishing message and said she was honored to do so (I think she had also fallen victim to it). A brilliant graduate student I had met that summer posted this in response: "Yeah, my first thought was 'wendy chun thought of me!!' Then my heart sank a lot, then I realized it was spam.:)." In response, I posted: "perhaps this is the upside of spam—contacting everyone with love for me." Although I was half joking at the time, there is something to the idea of spam as love: this exchange led to my thinking through the relation between Povinelli's discussion of the physical sores that mark contact in impoverished areas of Australia to virtual sores that are allegedly tied to "emerging" nations and markets. Both, that is, create "attitudes of interest and disinterest, anxiety and dread, fault and innocence about certain lives, bodies and voices and, in the process, form and deform lives, bodies, and voices."[65]

This loving side of spam also undermines the difference between spam and not-spam, human and inhuman. After all, what is the difference between semiautomatic "happy birthday" postings on Facebook pages and the emails, allegedly from friends, asking users to buy drugs from dodgy Canadian pharmacies? Involuntary (or not entirely voluntary) messages from others remind users that they are somehow connected to others, that they are in their address book, that others care enough about them to put them at risk. Also, as the founder of Slashdot, Rob Malda noted, slashdotting a site often makes it inoperable: a hug from a mob is indistinguishable from a distributed denial of service attack.[66] Again, moments of synchronous 'we,' of communal action, can destroy networks; YOUs value has the power to undo itself.

These interactions remind us that freedom and friendship are experiences that deny subjectivity, as much as they make it possible. As experiences, they are not contractual, but rather are perilous efforts, and we do not know in advance where they will lead. As Jean-Luc Nancy has argued, freedom is an experience. It is "an attempt executed without reserve, given over to the *peril* of its own lack of foundation and security in the 'object' of which it is not the subject but instead the passion, exposed like the pirate (peirātēs) who freely tries his luck on the high seas."[67] The Greek root for "pirate" is also the root for both "peril" and "experience."

Friendship's freedom comes without guarantees. Further, it is not a thing we possess, not something that anyone can own or grant another, even if it generates YOUs value that some can temporarily capture. It is a force that breaks bonds: a form of destruction that, Nancy argues, enables both friendship (habitation) and total destruction. Friendship as an experience is a moment of both terror and hope: a moment of hosting without meaning to and of being hostage to the other.

I Never Remember; YOUs Never Forget

=start unafraid and documented?

On April 5, 2013, a series of undocumented young adults posted videos to *Youtube.com* in support of the U.S. DREAM Act, a legislative act to grant resident status to persons of "good moral character" who entered the United States before the age of sixteen. In these videos, individuals 'came out' as undocumented and unafraid; they relayed stories of their personal struggles and their demands for justice. In one of the most viewed videos (10,183 views as of October 22, 2015), Maria Marroquin states:

> My name is Maria and I am undocumented. If you are watching this video, it is because I've been arrested.
> [Maria Marroquin
> Undocumented and Unafraid
> Pennsylvania
> 23 yrs old]

I was born in Lima Peru and I came to this country when I was thirteen years old, and since then I was enrolled as a ninth grader, and I studied hard, and I got good grades, but I realized I was different in my junior year in high school. I realized that because of my status, or my lack of status, I was not going to be able to continue with my education and go on to college. Despite all the obstacles, I decided to work hard, and I graduated from high school in 2004 with top grades, and then enrolled at a community college as an international student because currently in Pennsylvania undocumented students like me are forced to pay international

students' tuition, with no financial aid. It took me five
years to complete a two-year degree, and I graduated last
year in 2010 with an Associate's degree in social sciences
and a 3.98 GPA. Despite all these achievements, I still find
myself stuck because I cannot continue with my education to
become an immigration lawyer. Right now I am doing this
because I don't want my brother and my sister to go through
the same struggles I went through. I do not want them to go
through five years of school to complete a two-year degree.
[*Break … Sorry. Crying …*
I can cut all this out … look up.
Sorry.
Don't apologize; crying is the highest form of strength.]
I am doing this right now because in Georgia there are laws
and bans on students and I am tired of seeing students being
criminalized for wanting to obtain an education. I am tired
of seeing students lose hope because they cannot realize
their dreams of living freely in this country.[1]

As Cristina Beltrán reveals in her insightful analysis of
the work of the DREAM activists, these videos—and the
"Coming Out of the Shadows" campaign more generally—
consciously deploy tactics of visibility developed by
LGBTQ activists.[2] More confrontational and creative than
previous forms of undocumented immigrant activism, they
create "new spaces in which the undocumented are not
objectified members of a criminalized population who are
simply spoken about but instead are speaking subjects and
agents of change."[3] Although Marroquin's narrative
emphasizes her academic commitment, other activists use
"humor, anger and irony" to fight against their
criminalization.[4] In general, they refuse to apologize for
their or their parents' actions and fight against their
criminalization. Beltrán argues that radical DREAMers
"queer the movement, expressing more complex and
sophisticated conceptions of loyalty, legality, migration,
sexuality, and patriotism than those typically offered by
politicians, pundits, and other political elites."[5]
Specifically, they refuse to deploy a xenophilic

strategy—that is, one that celebrates immigrants as the ideal outsider citizens—both because it has not been successful in the past and because it marks them as "forever foreign."[6] Although Beltrán is careful not to simply celebrate this queering, since doing so can fall into the trap of "homonationalism" and thus support narratives of American exceptionalism (i.e., Look, we Americans are so more enlightened than 'savage' countries, which do not respect the rights of homosexuals), she does see this queering—in particular the work of unapologetic "undocuqueer" activists—as transforming, rather than simply accommodating to, existing social structures.

These activists engage in risky activities—they expose themselves and thus court deportation—because they realize that privacy offers no shelter against surveillance and prosecution, just as claiming the position of the "ideal immigrant" does not lead to inclusion: for the undocumented, the "private realm serves as the site of a social order characterized by secrecy, exploitation, and fear."[7] This rejection of privacy—specifically the privacy protection of normative national culture—is, as Lauren Berlant has argued, quintessentially queer.[8] Further, by refusing to remain in the shadows and by making demands to authorities they do not entirely trust, these youth, as Beltrán notes, embrace what Bonnie Honig has called "gothic" notions of power.[9]

These protests are remarkable not only for their fearlessness, but also for their repetition. The lines "I am undocumented" and "I am unafraid" are reiterated over and over again, and each narrative follows a template. Each begins with the phrase, "My name is …" and is usually followed by the phrase, "I am undocumented." Like Marroquin's, Viridiana Martinez's cyber testimonial begins with: "My name is Viridiana Martinez. I am undocumented. If you're watching this video, I've been arrested."[10] Each activist then narrates how she or he entered the United States as a child, the struggle to stay in school, the desire for education, etc. (The DREAM Act promised

students with postsecondary education and/or those who
served in the U.S. military a path toward permanent
residency). Further, these DREAMers wear the same
T-shirts, featuring the phrases, "I AM UN-DOC-U-MENT-ED"
or "DREAM" or "QUIP" (Queer, Undocumented, Immigrant
Project). **They embrace and indeed accentuate *Youtube.com*'s
unrelenting template.** Rather than strike poses to mark
their uniqueness, they use camera angles, clothing, and
backgrounds to stress similarities; to create their own
documentation, or, to draw from the work of Sarah Banet-
Weiser, their own "brand" of authenticity.[11] They reveal
their personal truths—their secrets, however open (the
fact that they cannot register as in-state students
immediately marks them as "undocumented;" further, the
fact that most pay taxes and thus have Taxpayer IDs
demonstrates that they are "undocumented" in name only)—to
the public. In this sense they exemplify and occupy the
branding of authenticity that drives neoliberal modes of
empowerment.[12]

To understand the stakes of this 'coming out,' we need to
place it in the context of the general appropriation and
habituation of this gesture—especially in response to a
situation in which one is already 'outed.' In the
documentary *Citizenfour*, for instance, Edward Snowden
frames his decision to reveal his name to the public as a
form of coming out. In response to Glenn Greenwald's
question, "You're coming out because you want to fucking
come out?" rather than just because he thinks the
government already know that he is the leak, Snowden
replies:

I don't want to hide on this and skulk around, I don't think
I should have to ….
I think it is powerful to come out and be like, "Look, I'm
not afraid and I don't think other people should be either …"
I think that's brilliant. I mean,
your principles on this I love, I can't support them enough,
because it is, it's inverting the model … the government has
laid out, where people were trying to, you know, say the

truth, skulk around, and then hide in the dark, and then
quote anonymously. I say yes, fuck that.[13]

Russia's and China's embrace of Snowden is thus an odd
reversal of the usual mode of "homonationalism." Rather
than the United States congratulating itself for
sheltering those who are not safe elsewhere, Russia prides
itself on offering refuge to Snowden after his dangerous
'coming out.'

But what does it mean to confess in a medium, in which we
are always confessing? What does it mean to insist on
speaking when what one says is already known? That is,
what does it mean to 'come out' as a way to preempt the
inevitable? Intriguingly, the cinematography of
Citizenfour—with its constant emphasis on windows and
mirrors—visually relays the fact that Snowden is never
safe; the private is now public, we are always under
possible surveillance. So how can we understand this
impulse to expose our secrets—to authenticate ourselves—
when we are publicly exposed, even when enclosed in
private space? Lastly—and most importantly—how can we
understand the DREAMers embrace of templates as a means
of shelter and habitation?
=end unafraid and documented?

4   Inhabiting Writing: Against the Epistemology of Outing

Affect does not reside in an object or sign, but is an effect of the circulation between objects and signs. ... Signs increase in affective value as an effect of the movement between signs: the more signs circulate, the more affective they become.

—Sara Ahmed[1]

Instead of getting upset over the gigantic (or so they say) growth in our means of communication, and fearing through this the weakening of the message, we should rather rejoice over it, serenely: communication "itself" is infinite between finite beings. Provided these beings do not try to communicate to one another myths about their own infinity, for in such a case they instantly disconnect the communication. But communication takes place on the limit, or on the common limits where we are exposed and where it exposes us.

—Jean Luc Nancy[2]

Ignorance and opacity collude or compete with knowledge in mobilizing the flows of energy, desire, goods, meanings, persons.

—Eve Sedgwick[3]

On September 7, 2012, fifteen-year-old Amanda Todd posted a video, narrated through notecards, relaying her "never ending story" of online blackmail and on- and offline bullying, to *YouTube.com*. According to this video, it started rather innocently in the seventh grade when she began to frequent webcam chat sites with friends. Online, she received many compliments: she was called "stunning, beautiful" and asked to flash, which she eventually did. A year later, a blackmailing "capper" threatened to circulate her topless photo to everyone, if she did not put on three shows for him. A few months later, he posted her picture to a porn site and sent the link to almost everyone she knew, including her mother. Todd subsequently became depressed and suffered from panic attacks, eventually abusing alcohol and drugs. She changed schools to start afresh, but the blackmailer followed her, sending *Facebook.com* friend requests to her new peers from a

Figure 4.1
Still from Amanda Todd video, usually used as its icon

page that featured her breasts as its profile photo. Todd then began cutting and moved to yet another school, where she was, she relays, finally happy: isolated but happy. After sleeping with a friend, however, she was beaten by his girlfriend in front of her new school and peers. A video of her assault was posted, and her consequent suicide attempt by drinking Clorox mocked. Todd's video ends with the cards: "I have nobody / I need someone ☹"; "my name is Amanda Todd" (figure 4.1).

This video 'went viral' after her suicide and, one month later in December 2013, one of the many copies of the original video (the original was taken down shortly after her death) reached 16 million views.[4] This video was widely viewed as a cry for help, a tragic foreshadowing of her death, with the "I have nobody / I need someone ☹" still used as an icon for the video (figure 4.1); but Todd offered a very different interpretation in the post that accompanied her video:

Published on Sep 7, 2012

I'm struggling to stay in this world, because everything just touches me so deeply. I'm not doing this for attention. I'm doing this to be an inspiration and to show that I can be strong. I did things to myself to make pain go away, because I'd rather hurt myself then [sic] someone else. Haters are haters but please don't hate, although im [sic] sure I'll get them. I hope I can show you guys that everyone has a story, and everyones [sic] future will be bright one day, you just gotta pull through. I'm still here aren't I ?

—AmandaTodd (2012)[5]

For Todd, as her mother would later insist, this video documented her survival and her will to live; it encapsulated her desire to inspire and support others. It was not a suicide note.[6] In it, she retells, and reclaims, her story and her exposure. In it, her face is shaded and partly cut off, and the notecards are positioned strategically at chest level. We, the audience, are focused at the site of her initial exposure, but instead of seeing her breasts, we read her story. The notecards and her shadowy presence offer a shield and shelter, from which Todd voluntarily reveals her 'secret': not her story, but her name, Amanda Todd.

Using a notecard video to reveal a story of abuse (and later self-abuse) was not unique to Todd. Earlier, in March 2011, thirteen-year-old Alye Pollack had posted a video that used notecards, interspersed with images of her face, to narrate her story of being bullied and of her temptation to cut. She wrote: "Not a day has gone by without one of these words"/ "Bitch, Whore, Fat, Lesbo, Slut, Freak, Ugly, Weird, Fag" (figures 4.2 and 4.3).[7] Fourteen-year-old Jonah Mowry posted a similar video that August, in which he described how he was bullied because of his sexuality and how he subsequently began cutting himself. In December 2011, Mowry's video went 'viral,' garnering many million views and mainstream U.S. television news coverage.[8] Both Pollack's and Mowry's videos started with "Hello/Hi, I'm …" and cards that explained that their happy faces were lies (figures 4.4, 4.5, 4.6, 4.7, 4.8). This format—the revelation of a secret true and troubled

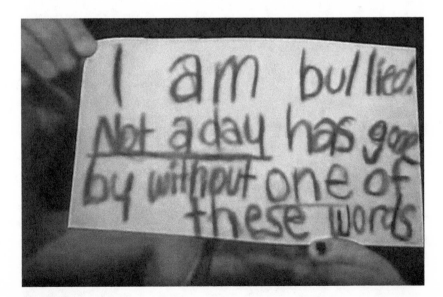

Figure 4.2
Still from Alye Pollack video

Figure 4.3
Still from Alye Pollack video

Figure 4.4
Still from Alye Pollack video

Figure 4.5
Still from Alye Pollack video

Figure 4.6
Still from Alye Pollack video

Figure 4.7
Still from Jonah Mowry video

Figure 4.8
Still from Jonah Mowry video

self—became standard in the many videos that would follow, and schools began to create notecard video assignments for students, as a way to combat teen suicide.[9] However, in addition to generating many positive responses (from national newspapers to thousands of videos and images produced in response), these videos engendered many negative ones. Within the comments sections of these videos were—and still are—hateful comments that accuse these people of being attention seekers, of being ugly, fat, lesbian, gay, of deserving the bullying they received, of deserving to die (see appendix). Although these comments are soon reported and deleted, their presence lives on in the many horrified responses to these comments, which are not erased.

How are we to understand and respond to these confessional videos and the reactions they elicit? How can we make sense of their rampant spread and the intense love and hate they continue to provoke? The answers to these questions are linked to the answers to following ones: **What does it mean to confess in a medium in which we are always confessing, to reveal secrets in a medium in which there are no secrets?** Why is shame the secret that is so often revealed in allegedly shameless social networks? What produces this voluntary exposure, especially around stories of bullying, sexuality, and self-harm? Why are these openly repetitive revelations of open secrets framed as revelations of a true self and thus also virulently attacked as inauthentic? Most pointedly: **What does it mean to condemn certain users as "whores" and attention seekers in a medium, a network, that is by definition promiscuous? AND, what does it mean to seek to find refuge—something like community—in a medium that exposes one to such hatred? Why write?**

To resolve these questions, this chapter demonstrates that these videos and their responses are part of a larger "epistemology of outing": a form of knowledge based on the forced exposure of open secrets, which has come to supplement what Eve Sedgwick diagnosed as the "epistemology of the closet" in 1990.[10] This mode of knowledge is also linked to the history of the confessional, outlined by Michel Foucault in *The History of Sexuality*, volume 1, a text against which Sedgwick argues. To be clear, by placing these videos and their responses within this conflicting and conflicted theoretical framework, I am not dismissing them as self-delusional.[11] Rather, I do so to break the seemingly unending cycle of judgment and condemnation of these videos based on the perceived 'authenticity' of their creators and to open different ways of understanding their power, their truth. This chapter thus concludes by drawing from the work of Jean-Luc Nancy to reveal how these videos struggle for something like community from the experience of the lack of community—or more properly, the experience of a community brought together through hatred. By writing, by embracing the template as a shield, these protagonists communicate by repeating—a message that we do not understand at all if we classify them as mere imitation or as 'viral.'

Outing Vengeance

Responses to the Todd and Mowry videos—the two notecard videos that received the most extensive broadcast news coverage in the early 2010s— have been (to some) shockingly mean-spirited. The first comment on December 9, 2013, to the Todd video, for instance, read: "drug addicted slut sucked off every boy drank bleach and killed herself haha hanging on a noose lol HAHAHAHAHAHA dumb bitch deserved death, lol your dead lol lol your dead" (see appendix). This comment is not anomalous; many comments insist "she deserved it" because she was "stupid" and because she was an "attention whore." To denigrate Mowry's video, comments focus more openly on its authenticity (presumably, since Todd committed suicide, her anguish was real, even if she was a "fake"). CNN Headline News's coverage of the Mowry's video featured the question: "Was Bullied Teen's Video Faked?"[12] This question responded to allegations based on a follow-up video in which a more cheerful and defiant Mowry (accompanied by a female friend) responded to his critics with: "to the people who say that no one likes me … uh … almost my entire school loves me … I don't want to sound stupid or conceited or anything … uh … PSS Thank you everyone who's being nice … and to the people who are being mean and calling me gay … um … thank you for stating the obvious. You could really be the next Einstein."[13] Although the "experts" on the *Dr. Drew* show concluded that Mowry's initial video was "real" (newsflash: teenagers are prone to mood swings), the fact that any trace of happiness or defiance could be perceived as undermining one's confession reveals the extent to which the 'success' of these notecard videos depends on a certain notion of 'pure' or 'true' victimhood, in which all observable actions must be pathetic, transparent, and consistent. Tellingly, Mowry's second video has since been deleted from *YouTube.com*.

This insistence on authenticity and transparency of character as the basis for true victimhood is also apparent in comments supporting Todd, which invariably wish death on the haters and/or emphasize Todd's beauty:

You, and those people who liked your comment … Sick twisted motherfuckers who shouldn't have been born. Why even write that? It's a shitty respectless opinion wich [sic] just pisses people off. Wanna be different or are you actually so fucked up that you really mean that??? Leave Earth please. thanks

...

it's so horrible; why would ANYONE do that; one mistake and everyone hurts someone so much. She was and [sic] angel, if any of you have seen that photo that said angels are the ones who harm themselves and can't take it; rest in peace to that very beautiful girl.

...

Why would people do this it is so messed up to bully a beautiful girl RIP Amanda Todd (Appendix)

As stated previously, this outpouring of adoration mimics the language Todd herself found so seductive: Todd exposed herself in response to compliments like the ones posted to her video after her death. As she states in the video: "In 7th grade I would go with friends on webcams" / "meet and talk to new people" / "Then got called stunning, beautiful, perfect, etc. ..." / "Then wanted me to flash."[14] The hostile messages posted by her supporters, in particular the repeated calls for "the haters" to kill themselves, similarly mimics the logic of Todd's initial postmortem tormentors, as well as Todd's own: "Haters are haters but please don't hate, although im [sic] sure I'll get them."[15] Even empathy is tinged with vengeance.

Not surprisingly, the responses to Todd's tormentors mirror the tactics of the tormentors themselves, namely the outing, or "doxing," of individuals. The reaction to Todd and other suicides has often been "human flesh search engines," through which perpetrators are found and punished.[16] As the Anonymous video posted in response to the Todd suicide and the vitriolic comments threatens/promises:

To the students and the adults who bullied the student Amanda Todd, we have a message for you. We know who you are. We know what you did. We are all around. We are your neighbors, your friends, your family. We know where you live. Take responsibility for your actions. As far as cyberbullies that posted derogatory comments on Amanda Todd's video, we will find out who you are. We will make you accountable. We do not forget. We do not forgive. Expect us. We are Anonymous.[17]

The price for vengeance: an even greater vengeance. The price for exposing others: an even greater exposure. No forgiveness, no forgetting. On both sides, once it's out there, it's there to stay; or, as Amanda Todd put it, "I can never get that photo back" (figures 4.9 and 4.10).

Although Anonymous's interventions are important, and anonymity is central to democracy—especially, as Danielle Allen has shown, to democracy in peril—is vengeance in exchange for vengeance, outing in response to outing, this logic of protection and forced accountability, the best

Figure 4.9
Still from Amanda Todd video

Figure 4.10
Still from Amanda Todd video

long-term solution?[18] Rather than punish those who punish others, should we not question why these indiscretions matter in the first place? Why should a topless picture drive a young girl to suicide? Further, given that we are all exposed online, given that our machines work by leaking, given how promiscuous our machines are, why this vitriol? Why do we hold people to a higher standard than our machines? **What work does blaming the user do? And what do we do with the increasing failure of shaming—the embrace of a certain shamelessness that is neither good nor bad?** If shame, as many psychologists have argued, is an internally produced emotion, what does it mean for shame to be a goal, rather than a response?[19]

Slut Shaming

Amanda Todd, among other things, was subjected to "slut shaming": the public shaming of women for their perceived promiscuity and their refusal to uphold the "zone of privacy" that shields normative heterosexual sex.[20] Slut shaming usually targets young women who embrace publicity as a means of empowerment. It responds to neoliberalism's rampant destruction of normative zones of privacy by seeking to create 'private' communities based on shaming.

Slut shaming exemplifies the dangers—and for some the pleasures—of Web 2.0 because of the logic of the example lodged within it. This exemplary logic involves condemning and/or sympathizing with the slut/victim and also spawns endless debates over personal responsibility—"Is the victim or those who circulated the images to blame?"—while eliding the question "Why should these images be considered harmful?" Further, it forestalls a rigorous engagement with networking and social protocols, for slut shaming transforms the consequences of the endemic publicity of the Internet (due to technological, social, and political infrastructures, as discussed in chapter 1) into user-induced accidents. **It blames the user—*her* habits of leaking—for systemic vulnerabilities, glossing over the ways in which our promiscuous machines routinely work through an alleged 'leaking' that undermines the separation of the personal from the networked.**

Slut shaming stems from "teachable moments." For example, in October 2011, the director of technology of the Georgia Fayette County Public School District held an assembly on Internet safety, during which he showed an image of then-seventeen-year-old Chelsea Chaney in a bikini standing next to a cardboard cutout of a rap star (figure 4.11).[21] Tagged with the line "Once It's There—It's There to Stay," this image (preceded by one in which a cartoon daughter confronts her mother for listing as her hobbies

Once It's There

It's There to Stay

- The Way Back Machine
1996-Present; All you need is
the URL

- Cached Pages
Google, Yahoo, Bing All
Search Engines Have Them

- Application Search
Facebook, YouTube,

Figure 4.11
Chelsea Chaney's disputed *Facebook.com* image

"bad boys, jello shooters, and body art") was taken without consent from Ms. Chaney's *Facebook.com* page. After school officials refused to apologize or hold a requested assembly on "respecting the rights of others on the Internet," Ms. Chaney sued the school district for two million dollars, arguing that the district had acted as a bully by branding her publicly as a "sexually-promiscuous abuser of alcohol"—that is, by essentially outing her as shameless and trying to shame her.[22] The moral of this and so many other stories of slut shaming is: don't be stupid enough to expose yourself because "Once It's There—It's There to Stay." **More insidiously, as Sarah Friedland has argued, the message is: once you've exposed yourself as a slut, as a consenting spectacle, as shameless, you deserve no protection, no privacy.** You can be exposed or shamed over and over again, especially as an example of a bad or stupid user.[23]

This ethos is also clear in "revenge porn" sites: websites that ostensibly post compromising images of former partners. (In actuality, tech repair people, among other folks, also supply these images.) These sites frame vengeance as betrayal, as outing: to avenge is to publicly circulate intimate photographs, once allegedly freely given, in order to express and generate contempt. (This response—anger or contempt rather than shame in response to contempt—is technically shameless and is, according to psychologist Sylvan Tomkins's argument, more normal).[24] Pointedly, several attempts to regulate revenge porn buttress the logic of slut shaming. For example, the 2013 law passed in California to protect victims of revenge porn sites did not extend to those who took "selfies." The idea, articulated below most bluntly by revenge porn site owner Hunter Moore, is that these women are too stupid to be protected:

You have these people who're making this law to protect the fucking retards who took the fucking pictures in the first place, who put that fucking brush in her pussy and sent it to some boy she met on the internet 15 fucking minutes ago and now she wants that guy to fucking go to jail because he put it up on the internet because he thinks it's fucking funny. It is fucking funny—you put a brush in your pussy, you stupid bitch, so why should we protect you? Think. Think, you stupid fucking whore. Think. Just think before you put that brush in your ass and take a picture.[25]

The assumption: consent once, circulate forever. Once you have put an image into circulation, it can be circulated forever. These women, like celebrities who deliberately cross the line dividing private from public, can be exposed shamelessly as shameless. Again, this logic blames the user, her habits of leaking, for the inherent leakiness of new media.

As mentioned previously, this linking of gender, race, and privacy has a long history within the United States, as Eden Osucha, Eva Cherniavsky, and Hortense Spillers among others have demonstrated. The right to privacy in the United States was defined in relation to a white femininity that was purportedly injured by the mass circulation images. This circulation was portrayed as intrinsically pornographic, as racialized and racializing. Osucha makes this point by comparing the case of Abigail M. Roberson, to which "The Right to Privacy" largely responded and which inspired New York State's first privacy law, and that of Nancy Green, the woman who would become "Aunt Jemima."[26] These women's images were used to sell related mass commodities: Roberson's to sell flour; Green's to sell pancake mix. Green's image—like that of so many people of color used in advertising during that period—was assumed to be general rather than individual; Green did not have—and was never perceived to have—a right to privacy. As Hortense Spillers has argued, captivity and slavery de-gendered the enslaved body, reducing the female slave body to unprotected flesh that could be exposed so that others could become private subjects.[27] Roberson, in contrast, was seen as "damaged" and profoundly shamed by the mass circulation of her image. Her body's image—and thus private bourgeois subjects in general—needed to be protected. Citing Cherniavsky, Osucha notes that historically, "white women's claim to a protected interiority receives the widest cultural sanction, insofar as white women are required to embody interiority *for others*."[28] This logic is still with us today: Justice Antonin Scalia, in the majority decision in *Kyllo v. United States*, which declared the use of heat-seeking technology on private homes unlawful (that is, until such technology became widely available), argued that this technology "might disclose ... at what hour each night the lady of the house takes her daily sauna and bath—a detail that many would consider 'intimate.'"[29] The term "lady" and the presumption of a "daily sauna and bath" make clear the class bias of this statement. Even more pointedly, in light of slut shaming, white privileged women are required to embody shame, so that 'we' might empathize with them; indeed so that a 'we' that empathizes may emerge.[30]

This logic, however, is complicated because this demand to embody interiority—that is, be shamed—is coupled with the simultaneous condemnation and celebrification of women who refuse to do so. In essence, we have the clash of two stereotypes: the proper young white bourgeois woman who is wounded and shamed by publicity, and the empowered young white woman who embodies publicity and consumption by "self-branding."

Empowered to Be Attacked

Sarah Banet-Weiser has most insightfully analyzed postfeminist modes of empowerment and authenticity through self-branding in *Authentic™: The Politics of Ambivalence in a Brand Culture*. Noting that, in the context of brand culture, an individual is "a flexible commodity that can be packaged, made, and remade—a commodity that gains value through self-empowerment,"[31] Banet-Weiser analyzes groundbreaking "girl entrepreneurs," such as Jennifer Ringley, Natalie Dylan, and Tila Tequila, all of whom used online exposure to cast themselves into influential and 'authentic' brands.[32] That is, they all engaged in practices of self-disclosure and transparency in order to become successful, and in so doing became models for young girls, such as Amanda Todd, who have come to see *Youtube.com* as a space in which to experiment and to be discovered. Noting that digital technologies sit at the intersection between public and private, Banet-Weiser argues that constructing oneself as an image entails "private disclosure—this is who 'I' am ... yet the 'I' that is created is marketed to a public of both known and anonymous subjects."[33] Although she refuses to distinguish between authentic and inauthentic modes of empowerment, Banet-Weiser does question the efficacy of this postfeminist mode of empowerment, which takes place within "preexisting gendered and racial scripts."[34]

The costs of this genre of empowerment are clear in the case of the mixed-race Amanda Todd. Todd regularly posted videos of her singing to *Youtube.com*. She performed for others on *BlogTV.com*, where she eventually flashed. Tellingly, the most damning comments directed toward her sought to undermine her authenticity: she was an "attention whore" and so deserved her fate; she was "fake." To fail in self-branding is devastating and demoralizing. Further, the vitriol directed at Todd for failing to be authentic reveals the self-righteous hostility directed against "young girls," a vitriol embodied in Tiqqun's critique of the "Young-Girl" as "the *model citizen* as redefined by consumer society since World War I."[35] Tiqqun, a French collective of authors and activists, describes the "Young-Girl" as a transparent, all-consuming nightmare, whose every action is carefully accounted for and directed toward self-valorization and beautification: she is the ultimate war machine of the Spectacle. Raised from the ranks of the dispossessed, she spreads consumerism and kills alterity, for she forces everyone else to submit—as she has—to the relentless logic of commodification. In language that strongly echoes celebrations of the "produser" (but with a negative valence), they write, "The Young-Girl is both production and a factor of production, that is, she is the consumer, the producer, the consumer of producers, and the producer of consumers."[36] By submitting to the "Young-Girl," 'we' become subject to the Spectacle and become commodities, and the pull of the "Young-Girl" is considerable, for she is herself literally money: something that circulates and makes things valuable. Everyone wants her. The end Tiqqun portray for the "Young-Girl," however, is not pretty: *"The Young-Girl prizes 'sincerity,' 'good-heartedness,' 'kindness,' 'simplicity,' 'frankness,' 'modesty,' and in general all of the virtues which, considered unilaterally, are synonymous with servitude. The Young-Girl lives in the illusion that liberty is found at the end of total submission to market 'Advertising.' But at the end of servitude there is nothing but old age and death."*[37]

Tiqqun's analysis, like slut shaming, blames the victim. It blames the "Young-Girl" for spreading the Spectacle everywhere. Their biting critique exemplifies the bitterness directed at "Young-Girls" for their refusal to be properly private, for their embrace of celebrity, and for their carefully manicured appeal, which refuses embrace and possession even as it spreads privatization. Tiqqun's supposedly progressive contempt for "Young-Girls" (a category that, they insist, is not really gendered) coincides with the ridicule that drives slut shaming and the mean-spirited comments posted on sites of would be "girl entrepreneurs." In all these cases, these women are condemned for being pathetic in their pursuit of adoration; they are exposed as exposed, as shameless. To understand the force of hostility directed at these young women, it is thus crucial to unpack the stakes of this exposure and this demand for privacy and shame. Shame, as Tomkins argues, implies a relation: the person who is shamed "still wishes to look at the other with interest or enjoyment, and to be looked upon with interest or enjoyment in a relationship of mutuality."[38] To shame someone is to insist that she or he still care.

The Epistemology of Outing

All the cases above exemplify what I have been calling the "epistemology of outing": a mode of knowledge production that focuses on the exposure of 'secrets,' open or not. Again, this epistemology of outing extends Sedgwick's analysis of the centrality of the homo/heterosexual divide to shaping twentieth-century society. In particular, it responds to Sedgwick's own thoughts on the limitations of the closet, namely her observation regarding her "relative inability, so far, to have new ideas about the substantive differences made by post-Stonewall imperatives to rupture or vacate" the closet.[39] The epistemology of outing engages this rupture of the closet, of the private spaces of secrecy: a rupture that is not simply liberating or good. To be outed, after all, is often a violent act, no matter how open the secret of sexuality may be. The goal of outing is shame (shame for being closeted, shame for what one does in the closet). It is no surprise then that most of notecard videos relate to episodes in which one is exposed and labeled as sexually 'aberrant.'

This epistemology, however, extends beyond—encompasses, bleeds into—other forms of exposure that are not obviously related to sexuality. To be clear, this is not to say that sexuality is irrelevant; it is rather to see the logic of the outing (inside/outside) as structuring communication more broadly. As Sedgwick herself argues, the "now chronic modern crisis of homo/heterosexual definition has affected our culture through its ineffaceable marking particularly of the categories secrecy/disclosure, knowledge/ignorance, private/public, masculine/feminine, majority/

minority, innocence/initiation, natural/artificial, new/old, discipline/ terrorism, canonic/noncanonic, wholeness/decadence, urbane/provincial, domestic/foreign, health/illness, same/different, active/passive, in/out, cognition/paranoia, art/kitsch, utopia/apocalypse, sincerity/sentimentality, and voluntarity/addiction."[40] As stated earlier, fears about cyberporn framed it as rupturing the walls of the home. It was an unstoppable window, placed in the bedroom of "our" children that threatened to overwhelm them and 'us' with indecency. The epistemology of outing takes advantage of this rupture, which is not new to new media, but rather as Thomas Keenan has argued, is central to political theory and theories of subjectivity.[41] Most pointedly, the epistemology of outing depends on the illusion of privacy, which it must transgress. The blackmailing capper, in many ways, epitomizes this drive to reveal secrets and to seduce others to do things 'in secret,' while at the same time seeking to remain unknown.

Crucially, this logic is not outside the Spectacle, but part of it. Describing the role of phone phreaks and trolls, Gabriella Coleman in "Phreaks, Hackers, and Trolls: The Politics of Transgression and Spectacle," challenges the current vilification of trolls by pointing to their constant play/performance. At the same time, Coleman, citing the work of Dick Hebdige on subcultures, points out that phreaks, hackers, and trolls "make a 'spectacle' of themselves, respond[ing] to surveillance as if they were expecting it."[42] "[They] take some degree of pleasure performing the spectacle that is expected of them."[43] This "epistemology of outing" has strong ties to Enlightenment thinking, which Sedgwick has critiqued as paranoid. Indeed, it is paranoid theory's pleasurable mirror, for it is a pleasure that comes from taunting power. As Foucault puts it in his description of the sexualized spirals of power and resistance: "pleasure ... comes of exercising a power that questions, monitors, watches, spies, searches out, palpates, brings to light; and on the other hand, ... pleasure ... kindles at having to evade this power, flee from it, fool it, or travesty it. ... Power ... lets itself be invaded by the pleasure it is pursuing; and opposite it, power assert[s] itself in the pleasure of showing off, scandalizing, or resisting. Capture and seduction, confrontation and mutual reinforcement. ... These attractions, these evasions, these circular incitements have traced around bodies and sexes, not boundaries not to be crossed, but *perpetual spirals of power and pleasure*."[44] This relation seems to perfectly encapsulate the lulz: it is resistance as a form of showing off and scandalizing, which thrives off moral outrage. This resistance also mimics power by out-spying, monitoring, watching, and bringing to light, that is, doxing. Snowden's reaction in *Citizenfour*—his desire to be found and to expose himself, for director Laura Poitras to paint the target on his back and for him to 'come out' of hiding—is another turn of this spiral.

The epistemology of outing relies on the secret; on exposing what is secret and thus also reifying secrets, however open, as things to be outed. Jodi Dean, in *Publicity's Secret: How Technoculture Capitalizes on Democracy*, has outlined the relationship between Enlightenment, paranoia, pleasure, and exposure. Arguing that the drive toward publicity has produced not democracy but rather its opposite (a system of distrust and conspiracy), she argues that publicity is the ideology of technoculture, for it "makes today's communicative capitalism seem perfectly natural."[45] This naturalization of publicity depends on the secret, which makes this failure seem contingent rather than constitutive. The logic goes like this: publicity has failed to produce a democratic society, not because of systemic failures, but because there are still secrets out there to be unearthed.[46] The secret produces conspiring subjects and distrust, which in turn keep up the search for secrets.

Intriguingly, a response to this logic of "revealing, outing, and uncovering" is the celebrity subject: the subject that presumes to be known, for to be known is to matter.[47] The celebrity subject is an agent, perhaps the agent in the current public sphere: "in technoculture's flows of capital and entertainment, celebrities seem to be the only actors, the only persons who can act."[48] Unlike the conspiring subject, whose pleasure comes from interrogating and outing, the celebrity subject, according to Dean, is driven by drive: she gains pleasure from repetition.[49] The celebrity subject answers the question of the conspiring subject, who is "always seeking, always uncertain, never satisfied" with the answer: "everyone knows about me. I'm a star, a celebrity ... instead of searching for the secret, one posits oneself as the secret, as that which is known."[50] To return to the window as political theory discussed in "Privately Public: The Internet's Perverse Subjects," the conspiring and celebrity subject seem to be on either side of this window.[51] One appears to be staring out, constantly looking and theorizing; the other would be outside, providing the spectacle to be analyzed.

This agency of the celebrity-subject, however, is also "icky": it is a "scoop" that resonates with "tabloid elements of scandalous, obscene content."[52] That is, this circulation "stains" everything it touches because it makes everything that circulates trivial and everyone who actively seeks circulation pathetic.[53] For, as Dean asserts, "technoculture's unendingly circulating information stream casts doubt on the value of this content. This is the creepy part of making oneself seen: once we offer ourselves up, once we are displayed on the screens of technoculture, we are as trivial as everything else."[54] The celebrity subject, however pathetic, though, is not naïve. She knows that she is not really a celebrity, and this cynical agency is key, for "even if one knows that she isn't a celebrity, she acts as if she believes she were. The technologies believe for her."[55] The nontransparent technologies before her supplies the celebrity subject with an audience.

Although this dynamic between conspiring and celebrity subjects on either side of publicity's window seems to perfectly describe that between "doxers" and "Young-Girls," this analogy is fundamentally flawed, for **the traditional window is strikingly reversed: the public figure here is on the inside, rather than the outside—the light comes from within the private sphere.** The Spectacle emerges from what should be—what once was—closeted private space and moves outward. As mentioned previously, whereas Keenan's question "What comes through a window?" refers to the light that comes through the window into the home, the question here is reversed: What comes through the window to the outside?[56] This framing and direction are crucial, for the images are "icky" precisely because they are images allegedly produced in the shadows of privacy, hence the call for shame. Indeed, the example Dean offers of the celebrity—a parody of the MasterCard "Priceless" commercial series in which a photograph of an office party reveals a woman to be wearing no underwear (an early fictional example of slut shaming)—highlights that, in networks, it is the privatized subject who acts.[57] Further, the would-be celebrity does not simply act; she is caught—outed—acting in public.

This outing complicates Dean's separation of the celebrity subject from the wounded subject. Dean writes that she "disagree[s] ... with the current preoccupation with wounded, victimized or infantile citizens. ... These accounts miss the way the subject of technoculture is the extreme realization of an ideal of publicity, a literalization of the notion of the actor in the public sphere."[58] But to what extent can these be separated, given the twinning of celebrity and shame? Further, given that networks work by leaking, to what extent is this 'outed' celebrity reviled precisely because he or she encapsulates the networked condition, in particular the simultaneous embrace and denial of two-way communications, which makes our use of computers seem personal and private?

Importantly, Tomkins links shame to two-way face-to-face encounters. Shame is linked to the dropping of one's eyelids or averting one's gaze in the face of an other whom one finds interesting or enjoyable.[59] Shame is an interruption of communication, for "by dropping his eyes, his eyelids, his head, and sometimes the whole upper part of his body, the individual calls a halt to looking at another person, particularly the other person's face, and to the other person's looking at him, particularly at his face."[60] The face is key here because the "eyes both receive and send messages of all affects and thereby increase the ambivalence about looking and being looked at."[61] Shame is thus fundamentally linked to an immediate situation of two-way communication, something seemingly denied while in actuality amplified in networked communications. The promiscuity and constant back-and-forth of our networks are denied by interfaces that seem to mirror, or promise limited, nonvisual forms of communication (unless directly requested). The lack of a face-to-face encounter does not mean that there is no communication, although it does mean that there is a fundamental disconnect between physical and networked seclusion. However much the act of surfing the web in the privacy of one's bedroom may resemble reading a print book, it is fundamentally different. Again, if D. A. Miller could point out the disparity between readers and characters by arguing that characters in fiction cannot look back, and that "the novel-reading subject … is never seen in turn, invisible both to himself (he is reading a novel) and to others (he is reading it in private)," the networked subject—on either side of the screen—is always seen, never invisible.[62] Or, to relate it to Baudrillard's celebration of the silence of the masses in the face of mass media, online subjects—YOUs—can never be silent; they can never resist by refusing subjectivity.

In this context, the notecard videos are remarkable, for the secret they reveal is not simply their identity or their belief in themselves as the secret. Although most of the videos are about sexuality, these notecard videos do not reveal whether the young adults making them are really gay; as Mowry states, in the case of his video, this is open knowledge. Instead, these videos normally reveal the secret of self-abuse in response to being outed, taunted, and ridiculed. In Mowry's video, he reveals that he too is ashamed, that he hates and cuts himself, that he is afraid of going to high school. What is confessed to is shame—a looking away—in response to being 'outed.' Shame is the secret that frames an encounter between those who seem shameless: one 'comes out' to one's shame.

Shameful Shame

Given that we are always exposed online, what does it mean to confess in a medium in which we are already confessing? What does it mean to relay one's story of suffering in a notecard video? On one level, these videos seem misguided and hardly emancipatory. As Foucault has shown, the confessional has a long and dubious history within the West, and the secret—in particular sexuality as the secret of the self—has been foundational to a certain will to power. "Western man," Foucault contends, "has become a confessing animal," so inoculated to the call to confess that we "no longer perceive it as the effect of a power that constrains us; on the contrary, it seems to us that truth, lodged in our most secret nature, 'demands' only to surface."[63] Noting the long history of the role of confession to operations of power, Foucault also states:

One has to be completely taken in by this internal ruse of confession in order to attribute a fundamental role to censorship, to taboos regarding speaking and thinking; one has to have an inverted image of power in order to believe that all these voices which have spoken so long in our civilization—repeating the formidable injunction to tell what one is and what one does, what one recollects and what one has forgotten, what one is thinking and what one thinks he is not thinking—are speaking to us of freedom.[64]

The celebration of the 'bravery' of notecard videos would seem to buttress this interpretation: they call on victims to reveal their secrets, not to censor themselves. And it is the enigma of sexuality as the secret that must constantly be confessed, that must constantly be discussed even when not explicitly mentioned. "What is peculiar to modern societies" Foucault contends, "is not that they consigned sex to a shadow existence, but that they dedicated themselves to speaking of it *ad infinitum*, while exploiting it as *the* secret."[65]

To dismiss these videos as just another page in the *History of Sexuality*, however, is not only to engage in what Sedgwick calls paranoid theory, it is also not to engage with these videos. To repeat, these confessions are not about sexuality per se, but rather about being abused, or, more properly, abusing oneself in response to abuse. What is confessed to and is allegedly liberating, is not the speaking of one's sexuality, but rather of one's wounding and shame. Most succinctly: what is sought is a release from shame by admitting to shame.

Shaming Community

Shame has an odd status in relation to social media. Shame is perhaps the thing that can't be named in a shameless society, because it speaks to a desire to connect in the face of contempt and self-loathing. Shame is the experience of subjectivity as the consequence of interrupted intersubjectivity. Shame strikes "deepest into the heart of man" because it is self-generated. Whereas terror and distress may hurt, they are wounds inflicted

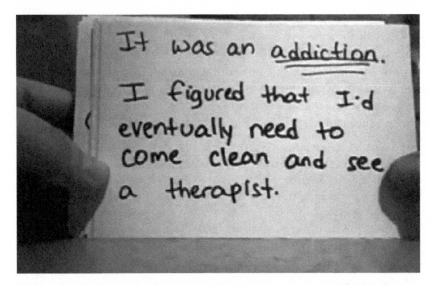

Figure 4.12
Still from video responding to Amanda Todd's ("Amanda Todd Suicide Response—
My Own Story," *Youtube.com*, October 14, 2012, https://www.youtube.com/
watch?v=EC0MOiOCcWA)

from the outside; in contrast, "shame is felt as an inner torment, a sickness
of the soul. It does not matter whether the humiliated one has been shamed
by derisive laughter or whether he mocks himself. In either event he feels
himself naked, defeated, alienated, lacking in dignity or worth."[66] Not surprisingly, the protagonist in a video posted anonymously in response to
Todd's (which takes the removal of the body of the speaker even further, by
exposing only hands) describes her online flirting as an "addiction": "I was
a straight A student. I was well rounded. No one suspected that I had a
secret like this. And I couldn't stop. It was an *addiction*. I figured that I'd
eventually need to come clean and see a therapist" (figure 4.12). The protagonist's addiction stems from the fact that shame is also linked to the
positive affects of love and identification.[67] The other who invokes shame
must also be an actual or potential source of positive affect, which survives
the expression of the other's contempt for the self.[68] That is, although both
shame and disgust "operate only after interest or enjoyment has been activated," shame incompletely inhibits this interest or joy.[69] Intriguingly,
Tomkins links shame to democratic societies and contempt to hierarchical
ones. "In a democratic society," he writes, "contempt will often be replaced
by empathic shame, in which the critic hangs his head in shame at what

the other has done; or by distress in which the critic expresses his suffering at what the other has done, or by anger in which the critic seeks redress for the wrongs committed by the other."[70] Shared shame inspires community, for "when one is ashamed of the other, that other is not only forced into shame but he is also reminded that the other is sufficiently concerned positively as well as negatively to feel ashamed of and for the other."[71] Shame produces shameful community.

The community that shame inspires, however, is not simply good. As Saidiya Hartman as shown in her remarkable *Scenes of Subjection*, empathy is hardly nonhierarchical. Slut shaming can be seen as an attempt to create community—decent community—by attempting to make certain women objects of shared shame.[72] More precisely, in slut shaming, the subject is punished for not enabling empathetic identification, hence the disturbing twinning of the vitriol directed at "sluts" and the empathetic celebration of true victims, where true = dead (and there is also the related twinning of the suicidal girl and the angry, vengeance-seeking boy, whose reaction to contempt is sometimes heroic action, sometimes persecution).

Sara Ahmed has, most presciently, addressed the emergence of community through hatred in *The Cultural Politics of Emotion*. Ahmed argues that hatred generates community—or more properly, communities (of the hated and the loved)—by aligning the particular with the general: I hate you because (you are a slut, etc.). Hatred as an "investment" "projects all that is undesirable onto another, while concealing any traces of that projection, so that the other comes to appear as a being with a life of its own."[73] This projection is also a bond between the hating subject and hated object, through which the subject emerges as a subject of love. As Ahmed puts it:

what is at stake in the intensity of hate as a negative attachment to others is how hate creates the 'I' and the 'we' as utter-able simultaneously in a moment of alignment. At one level, we can see that an 'I' that declares itself as hating an other (and who might or might not act in accordance with the declaration) comes into existence by also declaring its love for that which is threatened by this imagined other (the nation, the community and so on). But at another level, we need to investigate the 'we' as the effect of the attachment itself; such a subject becomes not only attached to a 'we', but the 'we' is what is affected by the attachment the subject has to itself and to its loved others.[74]

The 'we'—as communal and thus as loving—emerges through a hatred of the other, which aligns both the subject who hates and the despised object with others. Those who hate excessively need their objects, because they become part of a community through this attachment. This hatred organizes bodies and spaces. Hate is registered on the skin: through hate, the skin becomes a collective border: "It is through how others impress upon us that the skin of the collective begins to take shape."[75]

We need to understand hatred as central to community online—especially in relation to the vitriolic comments posted there. Hatred and trolling are not unfortunate sources of noise but rather signals or messages, as Lisa Nakamura has argued.[76] By seeing this, we can see how sites like 4Chan inspire and make possible communities, not only of trolls but also of those who love these dead girls, such as Amanda Todd. This excessive love, which depends on the valorization of the girls as "angels," seems out of proportion with the commenter's actual knowledge of these young adults. But it is here that the analogy and analysis falls short. Ahmed examines hate through the rubric of white supremacy, through those who frame their hatred as love. How, though, to understand this hatred in terms of a community of shame that still sees the victim as somehow the cause, rather than the effect, of networked vulnerabilities? Bizarrely, these women's actions are condemned most when they resonate with our machines' operations: when they reveal the ways in which we've been commodified and sold, precisely at the moments when we think we are in private. Indeed, these actions and these videos intersect, buttress, and call into question the presumption of privacy, and thus the logic of outing, so central to these videos.

So: to ask the obvious, how might we occupy networks differently? How can we understand publicity not in terms of a need for safety and protection, which is neither safe nor protecting, but rather **the fight for a space in which one can be vulnerable and not attacked?**

Anonymity, or the Right to Loiter

A more positive reading of the prevalence and deep resonance of these cases—a resonance also evidenced by the increasingly popular use of the term "raped"—would be this: these cases point to the fact that we need to fight for the right to be vulnerable—to be in public—and not attacked. We need to grapple with the ways that trust and publicity have always engaged risks. This was the message clearly delivered by the famous slut walks that started in Toronto in response to a Toronto police officer's implicit blaming of rape victims for the violence directed toward them.[77]

As Shilpa Phadke, Sameera Khan, and Shilpa Ranade have argued in *Why Loiter? Women and Risk on Mumbai Streets*, rather than fight for privacy—for hermeneutic bubbles of protection—we need to fight for the right to loiter.

As they put it, "we believe that it is only by claiming the right to risk that women can truly claim citizenship. To do this, we need to redefine our understanding of violence in relation to public space—to see ... the denial of access to public space as the worst possible outcome for women. Instead of safety, what women would then seek is the right to take risks, placing the claim to public space in the discourse of rights rather than protectionism."[78] As they note, this call for safety frames these women as "not safe in public spaces, sanctioning even greater restrictions on their movements."[79] Further, Phadke, Khan, and Ranade link the policing of women's bodies on Mumbai's streets to the policing of those of Muslim and other allegedly "dangerous men."[80] The logic here is perverse: in order to make public space safe for middle-class women, we need to remove both these women and 'dubious' men from these spaces, so that public spaces become the reserve of privileged men.[81] They thus insist that "women's open access to public space then cannot be sought at the cost of the exclusion of anyone else,"[82] for "it is only when the city belongs to everyone that it can ever belong to all women."[83] In order for this to happen, the authors contend, the urban infrastructure needs to change. Decent public toilets for women, well-lit areas, etc.—that is, places that welcome women, rather than assert their presence as a transgression—are needed. They also insist that these spaces be truly public, for "new spaces of consumption like coffee shops and malls are not public spaces, but privatized spaces that masquerade as public spaces."[84] These new spaces of consumption, which are trumpeted as safe for women (middle-class women with money), are not the answer.

Hence, rather than embracing the image and rights of the 'good' woman who deserves protection, Phadke, Khan, and Ranade seek a space in which risk can be courted. They further argue that the right to take risks—to enjoy oneself in public—is linked not to familiarity (indeed, the streets on which women are most recognizable are often the most dangerous), but rather to anonymity. They argue for loitering because mass loitering creates mixtures and possibilities that erode boundaries and create new spaces that do not 'leak' because boundaries are not compromised (and thus buttressed), they are fundamentally changed:

Loitering is significant because it blurs these boundaries—the supposedly dangerous look less threatening, the ostensibly vulnerable don't look helpless enough. *What if* there were mass loitering by hip collegians and sex workers, dalit professors and lesbian lawyers, nursing mother and taporis, [etc.] ... [T]his scenario might seem anarchic, but within this apparent chaos lies the possibility of imagining and creating a space without such hierarchies or boundaries.[85]

What might such a loitering online look like? How can we translate the call for mixture and better physical infrastructures to networks?

To loiter online, we would have to create technologies that acknowledge, rather than make invisible, the multitude of exchanges that take place around us—technologies that refuse the illusory boundary between audience and spectacle, author and character.[86] In addition, we would also have to build ones that question the basic premise that memory should equal storage, that everything read in should be written forward. Importantly, loitering is ephemeral: it inhabits the present. It also can transform 'open' private spaces into truly public ones.

Most importantly, we would need to engage in a politics of fore-giveness and deletion, in which we remember that to delete is not to forget, but rather to open other less dogmatically consensual ways of remembering. To fore-give is to give in excess, to give away—to create give in the system by giving way, by giving more than what one gets. That is, to build an Internet that embraces its status as a public domain, in which there is no promiscuous mode because there is no monogamous mode, we need to inhabit networks differently. We must develop new habits of connecting that disrupt the reduction of our interactions into network diagrams that can be tracked and traced. Our networks operate by fore-giving: signals, some of which we can read, are constantly caressing us. This mass touching—this mass writing—grounds communication.

Toward a Politics of Fore-giving

Writing touches upon bodies along the absolute limit separating the sense of the one from the skin and nerves of the other. Nothing gets through, which is why it touches.

—Jean-Luc Nancy[87]

To conclude, I want to return to the notecard videos, and in particular Amanda Todd's. These videos raise the questions: What does it mean to take on—to repeat—an action of exposure as a way to touch—to communicate with—others? How can one understand the outpourings they produced? Again, in these videos their protagonists, rather than confessing to being straight or gay or to having exposed themselves or not, relay the aftermath of having been outed: of having been exposed as exposing. Further, the secret they confess to is their vulnerability, their unhappiness, and their shame, although this is certainly not all they are 'about.' Indeed, rather than focusing on what they 'say,' I want to end by considering how they communicate and how they seek to create something like community, in the face of a community fostered by hatred.

Amanda Todd's video features a shadowy presence that seeks refuge by inhabiting rather than fleeing from the Internet. Instead of following police recommendations to get off the Internet if she wanted to avoid harassment (which is analogous to telling women not to dress provocatively or go out into public, if they want to avoid being raped), she posted this notecard video. If this video is about shame, or reveals shame, it does so not only through its narrative, but also through its form. If, as Tomkins argues, shame is fundamentally about the face—it is about a losing of face that nonetheless heightens the visibility of the face—notecard videos, by obstructing the face, enact shame. At the same time, because these videos still seek to communicate, they highlight the fact that shame is also a form of attachment. In shame, enjoyment and interest are interrupted, but they are still present as a desire: again, "in shame I wish to continue to look and to be looked at, but I also do not wish to do so."[88] Further, as Tomkins puts it, "shame is both an interruption and a further impediment to communication, which is itself communicated. When one hangs one's head or drops one's eyelids or averts one's gaze, one has communicated one's shame and both the face and the self unwittingly become more visible, to the self and others."[89] What is communicated is the desire to communicate in the face of inhibition.

Todd's video is shot at chest level, resolutely avoiding any face-to-face contact. Unlike in Mowry's and Pollack's videos, there is no moment in which the face/truth is revealed, either through a fake smile or a grimace indicating suffering. This refusal presumably protects Todd's anonymity, but, most importantly, it averts both the viewer's and Todd's gazes: we are both ashamed. The viewer's gaze though, is not simply averted; it is positioned at Todd's chest, at the place of her initial exposure. Instead of flashing her breasts, she narrates her story using flash cards. The writing is carefully timed, and she uses her body to create emphases by placing certain cards closer to the camera (figure 4.13). The misspellings and handwritten cards are odd in the age of spell-checking and glossy printouts—they mark her authenticity. What is most remarkable and makes this video different from the others is her open secret: her secret is not necessarily her suffering, which has been documented and ridiculed, but rather her name, Amanda Todd. She reveals what has been exposed over and over again; she gives up what once sought to hide. But this exposure of her name is exactly the point, for it is a gesture to inhabit this space, to refuse the ruse of privacy and to also assert her claim to be online. Through the communication

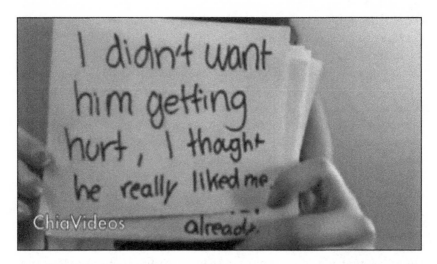

Figure 4.13
Still from Amanda Todd video

of shame—an interrupted communication that still communicates the wish to communicate—she asserts a continued shadowy presence. The video rejects the mask of anonymity even as, at the same time, it moves toward a presence that is not singular, but rather, to use Nancy's term, "singular plural."[90]

The Chia logo, attached to the most popular copy of the Amanda Todd video, is an odd manifestation of this. In a crass way, it places this singular experience (not only of suffering but also of suicide) within a larger series of Chia-branded videos. This branding seems disrespectful, especially because it implies that these experiences are all equivalent; but it also inadvertently reveals that this video is so compelling because it engages with branding, templates, and repetition. These notecard videos voluntarily inhabit templates, rather than accept them as necessary evils, which is the usual social media attitude (yes, every page is the same, but my content makes them different; it is my home page). These videos share the same style, sometimes the same story. **They relay their singular stories in a form that seems to deny singularity: repeating the narrative style, the notecard form, even the content. At stake in these videos and these outings or confessions is a reaching toward community, which stems from both what seems to be held in common but also what can never be: the singular experience of abuse and vulnerability.** That this form has been taken up as a way for victims of sexual assault to repeat phrases of abuse and for the 99% to tell their stories speaks to the ways in which the templated confessional has become central to questions of justice and redress.[91] Indeed, these videos' embrace of the template as the way to negotiate the demand to be individual makes them particularly interesting in this era of neoliberal empowerment and individualism. In this sense, they resonate with Baudrillard's reading of mass silence as a mass rejection of subjectivity.

Perhaps. At the same time, though, the notecard videos also seek to occupy a 'we': they reveal that even at the moment in which one feels most alone, one is always with another. This exposure—this repetition—reveals that one is never alone. At their best, the videos play with the singular plural that is the YOU. They inhabit it in order to produce a 'we' that does not flatten or align identity, but rather that reveals that singularity is fundamentally plural. Against the forms of community of hate/love referred to earlier, they seek community through exposure, for what is exposed, as Nancy argues, touches another. Importantly, this is not a question of virality: of one message infecting others, of communication as contagion. It is impossible to figure out which was the 'first' notecard video, as though there could be a first message; as though using placards for narration was native to new media. There is contiguity between these videos, but not continuity.[92] Their meaning is this 'we,' an originary multiplicity; the meaning of Being as communication, as "being-with-one-another."[93] It is a meaning that is not represented as society, but rather through writing.

Nancy has most rigorously theorized writing as communication, as repetition. Reading the work of Georges Bataille and addressing the repetition of writing at the end of writing, Nancy argues that writing "*exscribes*"— copies, disseminates—"meaning just as much as it inscribes signification."[94] Communication, he insists, is not about the communication of meaning or reasonable exchange: "it's not a question of that necessary, ridiculous machination of meaning which puts itself forward as it withdraws, or which puts on a mask as it signifies itself."[95] Rather, writing is a "*knowing nothing*" that "uses the work of meaning to expose, to lay bare the unusable, unexploitable, unintelligible and unfoundable *being* of being-in-the-world. *That there is* being, or some being or even beings, and in particular that 'there' is *us*, our community (of writing-reading): that is what instigates all possible meanings, that is what is the very place of meaning, but which has no meaning."[96] We repeat—we write, we read, we expose ourselves—to communicate this sense of community, to insist that this 'we,' however inoperative, however YOUs, is possible.

This repetition of writing that strains against meaning but that communicates a being-with also lies at the heart of so-called DDOS attacks. Started

by the Electronic Disturbance Theater in support of the Zapatista movement in Mexico, these computer-based performances used multiple (and
later automated) requests of a webpage to make that page unavailable, to
serve as a "virtual sit-in." Further, they used the "404 not found" error message to assert the absent presence of dead protestors, justice, truth, etc.
These protests were not about denial of service, but rather a collective "cry,"
a repetition that is the "renewal, rewriting of the petition, of the effort to
reach and join, of the request, of the demand, of the plea, of the claim, of
the supplication."[97] To see them as a denial of service attack is to change
perspective; it is to deny the fact that, as the Slashdot "hug" described in
the last chapter reveals, collective action—collective inhabitation—
overwhelms meaning. Networks operate through repetition. We are constantly caressed by signals that exscribe, that have everything to do with
communicating, but little to do with meaning. Networks work—they allow
us to communicate—by exposing YOU, by making YOU vulnerable, so that
there can be a 'we,' again however inoperative, however YOUs, to **begin
with.**

Indeed, this writing offers another way to understand the network maps
examined in chapter 1, which ground our current imagined networks (figure 4.14). This writing allows us to see that what matters most are not the
nodes and edges, but rather what happens, what touches at the edges. As
Nancy argues:

community is always beyond, that is, on the outside, offered outside of each singularity, and on this account always interrupted on the edge of the least one of
these singularities. Interruption occurs at the edge, or rather it constitutes the edge
where beings touch each other, expose themselves to each other and separate from

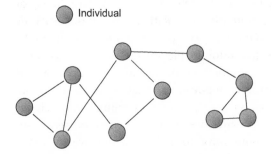

Figure 4.14
Social network representation, *Wikimedia*, January 2007, https://commons
.wikimedia.org/wiki/File:Social-network.svg by User: Wykis

one another, thus communicating and propagating their community. On this edge, destined to this edge and called forth by it, born of interruption, there is a passion.[98]

At the edge of exposure there is passion. And the network, if it is anything, is everything outside these nodes and edges. It is the everything that is the nothing communicated, that touches us—as we avert our eyes—as we hear and write.

=I have you in my pocket
According to several studies, a large majority of cell
phone users suffer from phantom vibrations: they feel
their cell phones vibrating when they are not. "Phantom
Vibration Syndrome" (PVS) was first 'diagnosed' by popular
media in articles such as the 2007 Associated Press piece
"Gadget Addicts Feel 'Phantom Vibrations,'" which Fox
News recirculated as "Cell Phone Junkies Feel Phantom
Ring Vibrations."[1] As these titles make clear, PVS was a
sign of addiction, a signal characteristic of a crackberry
addict. According to this article, PVS was allegedly:
• a sign that people feel they are "not whole" without
their cell phone;
• proof that, as human beings, "we're so tapped into our
community … / we're so attuned to the threat of isolation
and rejection, we'd rather make a mistake than miss a
call";
• evidence of extrasensory powers: one cell phone user
explained that the phantom vibrations were "anticipatory"
sensations. These users were "one with [their]
Blackberry."

Blackberries exemplify habitual new media: once deemed
crucial to the workings of corporate America, they were
displaced by the iPhone in the early 2010s. Nonetheless,
Blackberries linger through phantom vibrations and the
constant use they inspired: in a widely cited 2012 study,
89% of undergraduates reported experiencing "phantom
vibrations."[2]

Phantom vibrations also reveal how new media have changed
habit: habit has become addiction, and habituation is now
pathologized. Although framed as a form of "dependency,"
        most users do not experience phantom vibrations as
disturbing (some find it a point of pride).[3] For instance,
in the 2012 study, 40 percent of participants found the
vibrations to be not at all bothersome and 51 percent
found them to be a little bothersome. Only 2 percent of
users—those deemed most emotionally invested in text
messages—found them to be very bothersome. So, given the
lack of perceived harm or annoyance, why is this
experience even colloquially considered a "syndrome"?[4]

The name and symptoms of "Phantom Vibration Syndrome"
clearly draw from "Phantom Limb Syndrome," a widely
acknowledged disease in which amputees feel sensations—
pain, pleasure, etc.—from missing limbs. Extremely
difficult to treat, the cause of Phantom Limb Syndrome is
the subject of much controversy; but it has garnered much
recent popular attention, due to the use of "mirror
boxes," which enable some patients to alleviate pain by
allowing them to 'see' and thus 'unclench' their phantom
limbs.[5] The analogy between these 'syndromes,' based on
similar 'false' perceptions of sensations, also makes a
larger equivalence: it equates silent cell phones with
missing limbs. The upshot: our technologies have become
part of our body, and we are amputated without them.

Perhaps, but this easy lesson misses the fact that phantom
vibrations are not necessarily linked to missing cell
phones; these perceptions happen even when phones are
present. They are reactions to what one has, rather than
what one has lost. Further, the equation misses the ways
in which phantom vibrations indicate habituation. To
return to Ravaisson's definition of habit, habit is
receptivity become spontaneity. Rather than rely on actual
outside sensations, we spontaneously produce reactions,
and thus we satisfy ourselves—we no longer need cell
phones to provide 'good vibrations.'[6] So why are these
phantom vibrations pathologized? Indeed, more positively,

some of the medical literature frames phantom vibrations
as demonstrating the plasticity of the human brain.
According to this interpretation, phantom vibrations are
new perceptions rather than hallucinations.[7]

So: rather than trying to 'fix' them, how can we use these
phantom vibrations—in which we touch, or feel touched—as
grounds for, or revelations of, a more creative and
habitable future?

=end I have you in my pocket

Conclusion: Found Habituation

This book has argued that what matters most is what and how things linger. Rather than focus on the new and the fading—the bleeding edge of obsolescence—it has examined what remains, and how. Starting from a basic question about networks, namely, "Why have networks become the concept to explain everything new about our current era?," it has discovered habit (+crisis) as central to the temporality and logic of N(YOU) media. In imagined networks, connections are habits.

Habits, creative anticipations based on repetitions, ground network analyses, even if these analyses seek to explain what seems antithetical to habit: viral spread, crisis, and catastrophe. This is because habit itself is changing: it is increasingly understood as addiction (to have is now to lose). A significant number of popular and scholarly studies of habit treat it as something that needs to be changed, hence the coupling of habit with crisis. Now, to be is to be updated: Habit + Crisis = Update. This formula drives new media development. Further, through habit, networks are scaled. Habit allows us to move from individual tic to collective probabilities, for habits, however personal, are also things that collect. Habit makes possible YOUs value, "Big Data."

At the same time, this book has not simply condemned habit and N(YOU) media, but rather sought to comprehend the modes of inhabitation that they can shelter. Instead of mourning the loss of a 'we'/'they,' it has addressed habit as a way to inhabit the inhabitable. It has sought to understand the possibilities of YOU as singular-plural, as an entity that makes possible what Nancy has called an "inoperative community." Central to this effort has been the reframing of habit as publicity, as the remnants—or even scars—of others that one shelters within the self. Thus, to make explicit the larger argument driving this book (and the books that have preceded it): to address the pressing issues posed by the many networks around us, we need to focus on modes and modalities of publicity, instead of simply and constantly defending a privacy based on outdated notions of domesticity (privacy, that is, as house arrest). As many before me have argued, privacy and publicity are not opposites: privacy is a way of doing publicity. To be clear, this does not mean that privacy is not important, but rather that, in order to create a viable privacy, we need to grapple with—rather than cover over or ignore—the fundamentally intrusive nature of networks.

This book is a call for us to develop public rights, rather than accept the notion that if one is (un)wittingly exposed, one is then forever denied protection. Rather than "consent once, circulate forever," we need to find ways to loiter in public without being attacked. We need a politics of fore-giving that combats the politics of memory as storage, that fights for the ephemeral and fights not only for the right to be forgotten but also the right not to be stored in the first place. This reengagement with memory also entails a change in our habits of using—and our refusal of designs that undermine habituation by turning habits into forms of addiction, a refusal of undead information that renders us into zombies. It means inhabiting and discovering how our habits collect, rather than divide, us.

To conclude, I want to describe two projects that take on the complex relationship between private and public, individual and collective: Natalie Bookchin's *Mass Ornament* and *Testament*.[1] *Mass Ornament*, a single-channel video installation with five channels of sound, is a mass dance, constructed from clips of hundreds of vlogs. Starting with ambient noise and the mise-en-scène of these vlogs—private enclosed spaces such as bedrooms, living rooms, etc.—it then intercuts music from Busby Berkeley to Leni Riefenstahl films, as it sets the scene for the "chorus line" that follows (see figure 5.1).

Figure 5.1
Still from Natalie Bookchin's *Mass Ornament* (2009)

Crucially, before we get to the dancers in motion, screens and mirrors are used to emphasize the relationship between audience and vlogger. As Bookchin explains:

I edited videos of people dancing alone in their rooms, to create a mass dance reminiscent of historical representations of synchronized masses of bodies in formation, from Busby Berkeley to Leni Riefenstahl. I wanted the work to continually shift between depictions of masses and that of individuals. The dancers, alone in their rooms, seem to perform the same movements over and over as if scripted. But at the same time their bodies don't conform to mass ideals, and their sometimes awkward interpretations undermine the "mass ornament" produced by synchronizing their movements. I added sounds of bodies moving about in space, thumbing, banging and shuffling, as well as ambient sound emphasizing geographical differences, from crowded urban dwellings to the suburbs. Dancers push against walls and slide down doorways, as if attempting to break out of or beyond, the constraints of the rooms in which they seem to be encased.[2]

Mass Ornament captures 'our' neoliberal condition, in which 'we' are all allegedly individuals and in which the private (corporations, individual rights, etc.) seems to have triumphed over the public or social. Through a brilliant reworking of Siegfried Kracauer's reading of 1920s chorus lines as reflecting the logic of Fordism (neatly organized rows of dancers and viewers), Bookchin's rows of *Youtube.com* videos reveal private actions as forms of repetition. Focused on actions shot in the home, *Mass Ornament* is not simply a negative critique—hey, we're all the same—but also a hopeful revelation of an unconscious community or what Jaimie Baron has called "found collectivity," which we can trace through the mass archive.[3] The work reveals the traces of publicity that form users' habits and their seemingly private "revelations."

This found collectivity—which transforms the chorus line into an involuntary chorus of actions and voices—is also central to the poignant and insightful multichannel video installation *Testament*. *Testament* too draws from hundreds of vlogs, but rather than being set to a soundtrack of 1920s film, it uses the original words of the vloggers to create a chorus that addresses issues such as getting laid off, coming out, and taking medication. In this series—which makes the experience of the database strangely affecting—Bookchin uses instances of synchrony (such as moments in which everyone says "Xanax") to create magnetic moments that reverberate. Bookchin's project also focuses on dissonances (moments of outing) that reveal the uglier sides of communities (something further explored in *Now he's out in public and everyone can see*).[4] As Bookchin explains, the Greek chorus "embodies the reactions of audiences and the people against the kings and their misdeeds [in Greek dramas]. ... In the various chapters of *Testament*, I've created choruses of vloggers who comment on actions that have taken place off screen." As she notes, this is especially clear in *Laid Off*, in which she "compiled and edited together videos in which people discuss losing their jobs into a kind of talking choir. The actors, that is, those that have produced the tragedy—heads of companies, Wall Street, Alan Greenspan, our political system—are not heard from directly. Instead, we hear from a choir of 'the people' or 'the masses,' united in their language, as well as in their anger, frustration, and their despair over the economic crisis and its impact on their lives."[5] This chorus, though, is also dissonant: the members do not speak in unison, pointing to the complexities of individuality in collectivity.

The task before us is simple: to grapple with these moments, which are as much instances of found habituation as they are of found collectivity. And, through these moments, to try to inhabit the inhabitable, to give in excess and in advance, so that we can re-member differently.

Appendix: Selection of Comments Posted to Amanda Todd's Video (September 12, 2013)

1 hour ago

drug addicted slut sucked off every boy drank bleach and killed herself haha hanging on a noose lol HAHAHAHAHAHA dumb bitch deserved death, lol your dead lol lol your dead

Reply

15 hours ago

People bullyed me in my last name even know it looks funny it isnt whoever makes fun of it i hate them

Reply

12 hours ago

it is ok,Tyler you can get through it. IF ANYONE IS OUT THERE THAT IS GETTING BETTER TALK TO PEOPLE AND YOU WILL GET THROUGH IT

Reply

1 day ago

I AM FREAKING RAGING RIGHT NOW. FUCK THOSE BULLIES. FUCK EVERYTHING THEY'VE DONE.

3

1 day ago

I'm so sad :(((((R.I.P Amanda I love you

7

1 day ago

What happend to amanda todd was discusting i despise what them childish cowards did to her! Your all sick in the head commentin crap bout amanda go crawl back under the rock you came from!!! Sick twisted people

Reply

2

19 hours ago

Bullying girls are the worst!! Like this girlfriend of the guy who invited her for example! Bullies just bully to feel better because they have no self-confidence wothout doing that >:|

Reply

15 hours ago

She showed her tits like an idiot.

Reply

1

View all 6 replies

11 hours ago

+THEREPTILE666 YOU LITTLE PIECE OF SHIT, YOU SHOULD BE ASHAMED. HOW DARE YOU, EVERYONE MAKES MISTAKES.

Reply

11 hours ago

+Luke Templeton The difference is that it wasn't exactly a mistake, need to look up the definition? It was entirely on purpose, out of her own will.

Reply

13 hours ago

Noo , i want to meet you, i dont had the chance to talk to you WHY????

Reply

19 hours ago

These people are freaking sick they have no idea what they are doing. I mean what if they were them? They would probably do that to and I am sending a warning if I ever find the people who hurt her I will do exactly everything that they would do to her. And those people think they are awesome and famous cause they did that. But no they are immature stupid little fats. They are also not cool not popular not welcome and the worst people on earth. And if you are reading this comment give it a thumbs up of you think the people who hurt her feelings and who phys-icaly hurt her are little fat butts. I hope the people relize what they did killed her. So thumbs up this comment if you think those mean people are real ***** and pieces of **. And thumbs up this comment if you would have stood up for Amanda Todd and be her friend!!

Read more

Reply

1

1 day ago

Omg she seamed like a pretty and sweet girl to. She just made some mistakes and people shouldn't have wanted to die for those mistakes. RIP amanda

Reply

1

12 hours ago

Should had not shown her tits, what a whore

Reply

3 hours ago

A whore is where a woman haves sex more than one person and gets money off of the man kinda like a prostitue except not roaming the streets, so your comment is invalid...

Reply

1 day ago

she has me ;D f*ck everyone ;D haha no offense

Reply

9 hours ago

Bullies may be beautiful and handsome on the outside, but are the UGLIEST people on the inside.

Reply

6 hours ago

This is really sad not only of her story but this happened in my hometown as well.

Reply

9 hours ago

Harold pothead shipman fuck u so she showed her boobs that doesn't mean that everybody should just start hateing her fuck everybody who bullied her and fuck whoever says they r glad shes dead fuck u

Reply

6 hours ago
Shit
Reply

2 hours ago
When we cry god crys with us_

1 minute ago
Fuck alyuh hating bitches. ..I am being bullied rite nw .... I was beatin for
ah man day is nt even my mines.... I was cyber bullied by yhe girls dat
beat me >__< I dos cut my handa alot! I even try killin my self :'(I hp
wen I try again it relE wrks dis worl is nt fuh me .... :'(I hate living :'(
Reply

1 day ago
So she flashed people multiple times, posted pictures with her friends and
claimed she had no one, had sex with someone else's boyfriend and
couldn't suffer the consequences of her actions. I know people are all
like "she is just a kid" but if you post multiple nude pictures of yourself,
a picture of yourself giving someone a blowjob on Facebook, and have
sex, then you are no longer a child. The way I see it is she made terrible
choices and didn't learn from them because she made the same terrible
choices on multiple occasions and couldn't deal with the consequences.
This girl didn't deserve to die, but she doesn't deserve sympathy because
she was a coward, a slut in denial, and an attention whore.
Show less
Reply
3

22 hours ago
There is something very wrong with you...seek out some help.
Reply

20 hours ago
You, and those people who liked your comment... Sick twisted motherfuckers who shouldn't have been born. Why even write that? It's a shitty
respectless opinion wich just pisses people off. Wanna be different or are
you actually so fucked up that you really mean that??? Leave Earth
please. thanks
Reply

19 hours ago
The person who made her flash are pretty f***ed up
Reply
1

15 hours ago
omfg who would do that to someone. no one deserves that. they bullies
should have feel so bad.
Reply

17 hours ago
Amanda's bullies are awesome for making this bitch kill herself, yaaaaaay
XD
Reply

2 days ago
it's so horrible; why would ANYONE do that; one mistake and everyone
hurts someone so much. She was and angel, if any of you have seen that
photo that said angels are the ones who harm themselves and can't take
it; rest in peace to that very beautiful girl.
Reply

2 days ago
I understand it's upsetting she killed herself but teens kill themselves all the
time and don't get noticed this much what's so different about her???
Reply

2 days ago
Billy Nye what the ** is wrong with you how would you feel if you were in
this situation
Reply
2 days ago
All you fucking people that continu insulting her even though she is dead
you are fucked up saying stuff like that is stupid !!

2 days ago
STUPID HOE
Reply

3 days ago
And f#@k anybody that says something bad about this Amanda
Reply
1

2 days ago
To all those bullies out there, I wanna ask you a question. Are you perfect?
No. Of course not. No one is fucking perfect so don't go around telling
people they're sluts, bitches, losers, nerds etc. Don't fucking judge people by how they look! You're just telling them bullshit and most of them
will believe you! And what will that lead to? If you keep telling them
about the'r mistakes and that it will never get better, YOU WILL FUCK-
ING KILL THEM! Stop torturing innocent people! They may not be as
strong as they look. Stop the bullying! GET A LIFE!
Reply

2 days ago
I dont understand why there are people who watch this happen to such a
beautiful girl and dont do anything why!? R.I.P amanda todd no one will
ever forget you!
Reply

Notes

Introduction

1. Catherine Malabou, "Addiction and Grace: Preface to Félix Ravaisson's *Of Habit*," in Ravaisson, *Of Habit*, trans. Clare Carlisle and Mark Sinclair (London: Continuum, 2008), vii–xx. See Carlisle's and Sinclair's arguments for the "double law" of habit.

2. See Wendy Wood, Jennifer S. Labrecque, Pei-Ying Lin, and Dennis Rünger, "Habits in Dual-Process Models," in *Dual-Process Theories of the Social Mind*, ed. Jeffrey W. Sherman, Bertram Gawronski, and Yaacov Trope (New York: Guilford Press, 2014), 371–385; and Bas Verplanken and Wendy Wood, "Interventions to Break and Create Consumer Habits," *Journal of Public Policy and Marketing* 25, no. 1 (2006): 90–103. Verplanken and Wood contend, "Disrupting the environmental cues that trigger and maintain habit performance renders habits open to change" (91).

3. See David Harvey, *A Brief History of Neoliberalism* (Oxford: Oxford University Press, 2005); Naomi Klein, *Shock Doctrine: The Rise of Disaster Capitalism* (New York: Picador, 2007); and Lauren Berlant, *Cruel Optimism* (Durham: Duke University Press, 2011).

4. Milton Friedman, *Capitalism and Freedom* (Chicago: University of Chicago Press, 1982), xiv.

5. See Berlant, *Cruel Optimism*.

6. See David Liben-Nowell and Jon Kleinberg's description of email messages as spreading like a long thin chain. David Liben-Nowell and Jon Kleinberg, "Tracing Information Flow on a Global Scale Using Internet Chain-Letter Data," *Proceedings of the National Academy of Sciences* 105, no. 12 (2008): 4633–4638.

7. See Benedict Anderson, *Imagined Communities: Reflections on the Origin and Spread of Nationalism* (London: Verso, 1991).

8. See Gilbert Simondon, "The Genesis of the Individual," trans. Mark Cohen and Sanford Kwinter, in *Zone 6: Incorporations*, ed. Jonathan Crary and Sanford Kwinter (New York: Zone, 1992), 296–319.

9. For more on this, see Tony Bennett et al., "Habit and Habituation: Governance and the Social," in "Habit," special issue, *Body and Society* 19, nos. 2–3 (2013).

10. See Wendy Wood and David T. Neal, "A New Look at Habits and the Habit-Goal Interface," *Psychological Review* 114, no. 4 (2007): 843–863.

11. Wood et al., "Habits in Dual-Process Models."

12. Elizabeth Grosz, "Habit Today: Ravaisson, Bergson, Deleuze and Us," *Body and Society* 19, nos. 2–3 (2013): 219.

13. Wendy Wood and other psychologists who work on habit also express this sentiment.

14. See Malabou, "Addiction and Grace."

15. David Hume, *Treatise of Human Nature* (Oxford: Oxford University Press, 2000).

16. Maurice Merleau-Ponty, *Phenomenology of Perception*, trans. Colin Smith (New York: Routledge, 2002), 104.

17. Gilles Deleuze, *Empiricism and Subjectivity: An Essay on Hume's Human Nature*, trans. Constantin V. Boundas (New York: Columbia University Press, 2001), x.

18. See Clare Carlisle, "Creatures of Habit: The Problem and the Practice of Liberation," *Continental Philosophy Review* 38, nos. 1–2 (2006): 19–39, 29.

19. See William James, "Habit," in *The Principles of Psychology* (University of Michigan Library, 1890), http://psychclassics.asu.edu/James/Principles/prin4.htm.

20. Gabriel Tarde, *The Laws of Imitation*, trans. Elsie Clews Parson (New York: H. Holt, 1903).

21. Pierre Bourdieu, *Outline of a Theory of Practice*, trans. Richard Nice (Cambridge: Cambridge University Press, 1977), 78.

22. Ibid., 80–81. Bourdieu is quoting Gottfried Leibniz in this sentence. Bourdieu continues: "Habitus is precisely this immanent law, *lex insita*, laid down in each agent by his earliest upbringing, which is the precondition not only for the co-ordination of practices but also for practices of co-ordination, since the corrections and adjustments the agents themselves consciously carry out presuppose their mastery of a common code and since undertakings of collective mobilization cannot succeed without a minimum of concordance between the habitus of the mobilizing agents ... and the dispositions of those whose aspirations and world-view they express" (81).

23. James, "Habit," n.p. For more on the influence of William James on Bourdieu, see Gail Weiss's *Refiguring the Ordinary* (Bloomington: Indiana University Press, 2008).

24. James, "Habit," n.p.

25. Slavoj Žižek, *The Sublime Object of Ideology* (London: Verso, 1989), 28.

26. Pascal, quoted in ibid., 37.

27. Bourdieu, *Outline of a Theory of Practice*. Bourdieu argues, "if one regularly observes a very close correlation between the scientifically constructed *objective probabilities* … and *subjective aspirations* … this is not because agents consciously adjust their aspirations to an exact evaluation of their chances of success. … Unlike the estimation of probabilities which science constructs methodically on the basis of controlled experiments from data established according to precise rules, practical evaluation of the likelihood of the success of a given action in a given situation brings into play a whole body of wisdom, sayings, commonplaces, ethical precepts … and, at a deeper level, of the unconscious principles of the *ethos* which, being the product of a learning process dominated by a determinate type of objective regularities, determines 'reasonable' and 'unreasonable' conduct for every agent subjected to those regularities" (77).

28. See Geoffrey Hodgson, "The Ubiquity of Habits and Rules," *Cambridge Journal of Economics* 21, no. 6 (1997): 663–684.

29. See Grosz, "Habit Today"; also see Bourdieu, *Outline of a Theory of Practice*. Bourdieu argues that habitus, as history turned nature, reveals the extent to which yesterday's man predominates in us, albeit unconsciously as *habitus* (79). The habitus is the unconscious for Bourdieu.

30. See Charles Duhigg, *The Power of Habit: Why We Do What We Do in Life and Business* (New York: Random House, 2012), 14–15.

31. See Heidi Cooley, *Finding Augusta: Habits of Mobility and Governance in the Digital Era* (Lebanon, NH: University Press of New England, 2014).

32. John Dewey, *Human Nature and Conduct: An Introduction to Social Psychology* (New York: Henry Holt, 1922), 14.

33. Verplanken and Wood, "Interventions to Break and Create Consumer Habits," 91.

34. See ibid. and Ravaisson, *Of Habit*.

35. See Malabou, "Addiction and Grace."

36. See Margaret Thatcher in Douglas Keay, "Margaret Thatcher Interview," *Woman's Own* (1987), http://www.margaretthatcher.org/speeches/displaydocument.asp?docid=106689.

37. See "Most Highlighted Books of All Time," *Amazon.com*, n.d., accessed September 9, 2013, https://kindle.amazon.com/most_popular/books_by_popular_highlights_all_time. This page is no longer available online.

38. Ravaisson, *Of Habit*.

39. See Steven Shaviro, *Connected, or, What It Means to Live in the Network Society* (Minneapolis: University of Minnesota Press, 2003).

40. Harvey, *A Brief History of Neoliberalism*, 2.

41. As quoted in ibid., 23.

42. Michel Foucault, *The Birth of Biopolitics: Lectures at the Collège de France 1978–1979*, trans. Graham Burchell (New York: Palgrave Macmillan, 2008), 252.

43. Wendy Brown, *Undoing the Demos: Neoliberalism's Stealth Revolution* (Cambridge, MA: MIT Press, 2015), 17.

44. Foucault, *The Birth of Biopolitics*, 147.

45. See Harvey, "The Construction of Consent," in *A Brief History of Neoliberalism*, 39–63.

46. Friedman, *Capitalism and Freedom*, 4.

47. Ibid., 13, italics in the original.

48. Brown, *Undoing the Demos*, 36.

49. For two examples among many: Stephen R. Covey, *7 Habits of Highly Effective People* (New York: Free Press, 1989); Jennifer Cohen, "10 Morning Habits Successful People Swear By," *Forbes*, November 26, 2014, http://www.forbes.com/sites/jennifercohen/2014/11/26/10-morning-habits-successful-people-swear-by/.

50. Oskar Negt and Alexander Kluge, *Public Sphere and Experience: Toward an Analysis of the Bourgeois and Proletarian Public Sphere*, trans. Peter Labanyo and Assenka Oksiloff (Minneapolis: University of Minnesota Press, 1993), 18.

51. See John Stuart Mill, *On Liberty* (New York: Dover, 2002).

52. See Jürgen Habermas, *The Structural Transformation of the Public Sphere: An Inquiry into a Category of Bourgeois Society*, trans. Thomas Burger and Frederick Lawrence (Cambridge, MA: MIT Press, 1991); and Hannah Arendt, *The Human Condition* (Chicago: University of Chicago Press, 1998).

53. See Negt and Kluge, *Public Sphere and Experience*, 76; Hannah Arendt, "The Crisis in Education," in *Between Past and Future: Eight Exercises in Political Thought* (New York: Penguin, 1968), 186 and 173–196.

54. As cited in the Department of Justice brief (*Reno v. ACLU*), January 21, 1997, http://www.ciec.org/SC_appeal/970121_DOJ_brief.html.

55. See Ethan Zuckerman, "Cute Cat Theory Talk at Etech," ... *My Heart's in Accra*, March 8, 2008, http://www.ethanzuckerman.com/blog/2008/03/08/the-cute-cat-theory-talk-at-etech/.

56. Similarly compromised spaces include shopping malls, which are open yet privately owned spaces. Indeed, the public/private binary seems to have become transformed into one of open versus closed; open source software operates by spreading licensing everywhere, contributing to the demise of the public domain (see Wendy Hui Kyong Chun, *Control and Freedom: Power and Paranoia in the Age of Fiber Optics* [Cambridge, MA: MIT Press, 2006]). As Tiziana Terranova points out in *Network Culture*, corporations, which are treated as individuals in the United States, increasingly use strategies of "soft control," which encourage workers to view work as integral to their life and identity (see Tiziana Terranova, *Network Culture: Politics for the Information Age* [London: Pluto Press, 2004], 98–130; and see also Franco Berardi, *The Soul at Work: From Alienation to Autonomy* [Los Angeles: Semiotext(e), 2009]). Hype is central to the functioning of the global marketplace, in particular the stock market, in which the relationship between the price of stocks and corporate earnings is often tenuous, such as during the dotcom craze of the late 1990s. The importance of derivatives to the stock market further reveals the extent to which the possibility of change matters more than actual value. In a neoliberal economy, individualism is encouraged and allegedly valued: rather than being paid according to a scale, one is compensated according to one's alleged contributions and skills. The transformation of mass media into new media thus fits perfectly with this individualizing process.

57. For instance, the amount of sensitive information, from credit card numbers to corporate secrets, revealed publicly by cell phone users is legendary. The home phone and the public phone booth live on in the cone of silence unconsciously invoked when one talks "privately" on one's mobile phone.

58. See Ryan Singel, "Whistle-Blower Outs NSA Spy Room," *Wired*, April 7, 2006, http://archive.wired.com/science/discoveries/news/2006/04/70619.

59. Nicholas A. Christakis and James H. Fowler, "The Spread of Obesity in a Large Social Network over 32 Years," *New England Journal of Medicine* 357, no. 4 (2007): 370–379.

60. This study and its predecessor, the Framingham Heart Study, were initially set up to diagnose risk factors for heart disease. The Heart Study began in 1948 with a cohort of 5,209 people. The Offspring Study, which enrolled the original participants' children and their children's spouses, began in 1971 and had a cohort of 5,124 key participants (egos); it also included an additional 6,943 people (alters), connected in some way to the study's egos. The data for these studies is remarkably robust: only ten people left the Offspring Study, which offered the egos free physicals every three years. These data sets are also remarkably expansive, for they did not simply record health indicators such as weight, they also included "complete information about all first-order relatives (parents, spouses, siblings, and children), whether they are alive or dead, and at least one 'close friend' at each of the seven examinations" as well as detailed home addresses. Ibid., 372.

61. Ibid., 375–376.

62. Ibid., 376. What was even more surprising and not remarked upon is the role of gender: if a male alter became obese, the male ego's risk of obesity increased 100%; yet if a female alter became obese, the female ego's risk increased only 37%.

63. See Gina Kolata, "Catching Obesity from Friends May Not Be So Easy," *New York Times*, August 8, 2011, http://www.nytimes.com/2011/08/09/health/09network .html.

64. See Cosma Rohilla Shalizi and Andrew C. Thomas, "Homophily and Contagion Are Generically Confounded in Observational Social Network Studies," *Social Methods Research* 40, no. 2 (2012): 211–239; quotation from Hans Noel and Brendan Nyhan, "The 'Unfriending' Problem: The Consequences of Homophily in Friendship Retention for Causal Estimates of Social Influence," *Social Networks* 33, no. 3 (2011): 211.

65. See the Framingham Heart study website, https://www.framinghamheartstudy .org/about-fhs/index.php, updated 2015.

66. See Priscilla Wald, *Contagious: Cultures, Carriers, and the Outbreak Narrative* (Durham: Duke University Press, 2008).

67. Lisa Blackman, "Habit and Affect: Revitalizing a Forgotten History," *Body and Society* 19, nos. 2–3 (2013): 186–216.

Interlude: THEY→YOU

1. Martin Heidegger, *Being and Time: A Translation of Sein und Zeit*, trans. Joan Stambaugh (Albany: State University of New York Press, 1996), 123.

2. Ibid., italics in original.

3. Ibid., 163.

4. Ibid., 166.

5. Ibid., 168.

6. Daniel Paul Schreber, *Memoirs of My Nervous Illness*, trans. Ida Macalpine and Richard A. Hunter (1955; New York: New York Review of Books, 2000), xxi.

7. Jean Baudrillard, "The Masses: The Implosion of the Social in the Media," trans. Marie Maclean, *New Literary History* 16, no. 3 (Spring 1985): 578.

Part I

1. Bruno Latour, *Reassembling the Social: An Introduction to Actor-Network-Theory* (Oxford: Oxford University Press, 2005); Jean François Lyotard, *The Postmodern Condition: A Report on Knowledge*, trans. G. Bennington and B. Massumi (Minneapolis:

University of Minnesota Press, 1984), 15; Tiziana Terranova, *Network Culture: Politics for the Information Age* (London: Pluto Press, 2004).

2. Alexander Galloway, *Protocol: How Control Exists after Decentralization* (Cambridge, MA: MIT Press, 2004), 11.

3. This productive notion of the imagined has many sources: from Arjun Appadurai's description of the imagination as a key force in the era of "disorganized capital" in *Modernity at Large: Cultural Dimensions of Globalization* (Minneapolis: University of Minnesota Press, 1996), to Brian Keith Axel's notion of "the diasporic imaginary" in his essay "The Diasporic Imaginary," *Public Culture* 14, no. 2 (Spring 2002): 411–428, and Benedict Anderson's notion of the nation as an "imagined community," put forth in his book *Imagined Communities: Reflections on the Origin and Spread of Nationalism* (London: Verso, 1983). Networks also could be understood as the product of what Cornelius Castoriadis described as the "imaginary" in *The Imaginary Institution of Society* (Cambridge, MA: MIT Press, 1998). Importantly, though, networks are key to Castoriadis's own formulation of the imaginary, for his description of institutions presumes the existence of networks. The institution, he writes, "is a socially sanctioned, symbolic network" (ibid., iv). The network is key to the difficult linkage between the social-historical and the psychical, the collective and the individual, that the imaginary both enables and presupposes.

4. Anderson, *Imagined Communities*, 6.

5. Ibid.

6. Ibid., 7.

7. Ibid., 26.

8. For more on the notion of nearness and technologies' impact on it, see Martin Heidegger, *Being and Time: A Translation of Sein und Zeit,* trans. Joan Stambaugh (Albany: State University of New York Press, 1996), and "The Question Concerning Technology," in *The Question Concerning Technology and Other Essays* (New York: Harper and Row, 1977).

9. Anderson, *Imagined Communities*, 35.

10. Ibid., 26.

11. Allan Sekula, in "The Body and the Archive: The Use and Classification of Portrait Photography by the Police and Social Scientists in the Late 19th and Early 20th Centuries," *October* 39 (Winter 1986): 3–64, reveals the different archival logics that ground individual versus group identification in terms of photographs. Gilles Deleuze and Félix Guattari, in *A Thousand Plateaus: Capitalism and Schizophrenia,* trans. Brian Massumi (Minneapolis: University of Minnesota Press, 1987), influentially argue for the difference between molecular and molar groupings.

12. See Wendy Hui Kyong Chun, "Nómades que imaginan," in *Nomadismos tec-nológicos*, ed. Jorge La Ferla and Giselle Beiguelman, in English, Portuguese, and Spanish (Buenos Aires: Espacio Fundación Telefónica/Instituto Sergio Motta, 2011).

13. See Maurice Blanchot, *The Unavowable Community*, trans. Pierre Joris (Barrytown, NY: Station Hill Press, 1988).

Always Searching, Never Finding

1. Nick Carbone, "Top 10 Viral Videos," *Time.com*, December 4, 2012, http://entertainment.time.com/2012/12/04/top-10-arts-lists/slide/kony-2012/.

2. See Barbara Johnson, "Apostrophe, Animation, and Abortion," *Diacritics* 16, no. 1 (1986): 31.

3. The "we" is not the default: it is not assumed in an objective third-person narrative.

4. The notion of 'we' as temporally and spatially fixed corresponds to Émile Benveniste's claim that all personal pronouns "do not refer to a concept or to an individual" but rather to "the act of individual discourse in which it is pronounced." See Benveniste, *Problems in General Linguistics*, trans. Mary Elizabeth Meek (Coral Gables, FL: University of Miami Press, 1971), 226.

5. For more on this, see Thomas Keenan, "Publicity and Indifference (Sarajevo on Television)," *PMLA* 117, no. 1 (2002): 104–116.

Chapter 1

1. Arjun Appadurai, "Disjuncture and Difference in the Global Cultural Economy," in *Modernity at Large: Cultural Dimensions of Globalization* (Minneapolis: University of Minnesota Press, 1996), 31.

2. See David Harvey, *A Brief History of Neoliberalism* (Oxford: Oxford University Press, 2005).

3. Fredric Jameson, *Postmodernism, or the Cultural Logic of Late Capitalism* (Durham: Duke University Press, 1991), 83.

4. Fredric Jameson, "Cognitive Mapping," in *Marxism and the Interpretation of Culture*, ed. Cary Nelson and Lawrence Grossberg (Champaign: University of Illinois Press, 1990), 349.

5. Jameson, *Postmodernism*, 44. The increasing density of space and the waning of temporality foster this incapacity to map this great communicational network, which Jameson conjectured to be postmodernism's historically unique dilemma. That is, sounds and images relentlessly saturate space and make the world "a glossy

skin, a stereoscopic illusion, a rush of filmic images" (ibid., 34). Consequently, we, like schizophrenics, experience the world as a "rubble of distinct and unrelated signifiers" (ibid., 26). Faced with this breakdown of the signifying chain, we are incapable of cognitively mapping our relation to capitalist totality.

6. See Ulrich Beck, *Risk Society: Toward a New Modernity*, trans. Mark Ritter (London: Sage, 1992); originally published in German in 1986.

7. Ibid., 22.

8. Ibid., 72.

9. Ibid., 52.

10. Mark Granovetter, "The Strength of Weak Ties," *American Journal of Sociology* 78, no. 6 (1973): 1377.

11. See Kevin Lynch, *The Image of the City* (Cambridge, MA: MIT Press, 1960).

12. See Louis Althusser, *Lenin and Philosophy and Other Essays* (London: New Left Books, 1971), 162.

13. Jameson, *Postmodernism*, 39.

14. Ibid., 37–38.

15. Ibid., 37.

16. Gilles Deleuze and Félix Guattari, *A Thousand Plateaus: Capitalism and Schizophrenia*, trans. Brian Massumi (Minneapolis: University of Minnesota Press, 1987), 12.

17. Ibid.

18. Bruno Latour, *Reassembling the Social: An Introduction to Actor-Network-Theory* (Oxford: Oxford University Press, 2005), 128.

19. For more on affect as network, see Silvan Tomkins, *Shame and Its Sisters: A Silvan Tomkins Reader* (Durham: Duke University Press, 1995); Brian Massumi, *Parables of the Virtual: Movement, Affect, Sensation* (Durham: Duke University Press, 2002); and Patricia Clough, *Autoaffection: Unconscious Thought in the Age of Technology* (Minneapolis: University of Minnesota Press, 2000).

20. Mung Chiang, *Networked Life: 20 Questions and Answers* (Cambridge: Cambridge University Press, 2012), 8.

21. Ibid., 4. This notion that cellular behavior creates globally complex and efficient actions has a long history, of course, stemming at least from John von Neumann's early work on cellular automata.

22. Douglas Keay, "Margaret Thatcher Interview," *Woman's Own* (1987), http://www.margaretthatcher.org/speeches/displaydocument.asp?docid=106689.

23. See Alexander Galloway, *Protocol: How Control Exists after Decentralization* (Cambridge, MA: MIT Press, 2004); and Tiziana Terranova, *Network Culture: Politics for the Information Age* (London: Pluto Press, 2004).

24. Latour, *Reassembling the Social*, 9.

25. See Terranova, *Network Culture*, which moves us away from understanding networks, such as the Internet, in terms of infrastructure, and toward seeing them as flows of interactions. She argues for the importance of affective relations and for the need to create common passions that move across the informational milieu.

26. Ien Ang, "In the Realm of Uncertainty: The Global Village and Capitalist Postmodernity," in *Living Room Wars: Rethinking Media Audiences for a Postmodern World* (New York: Routledge, 1996), 162–180.

27. For more on the difference between these, see Michel de Certeau's *The Practice of Everyday Life*, trans. Steven Randall (Berkeley: University of California Press, 1984).

28. Duncan J. Watts, *Six Degrees: The Science of a Connected Age* (New York: Norton, 2004), 43.

29. Ibid.

30. Ibid., 29.

31. See "Fueling Long-Term Impact," *TeachforAmerica.com*, https://www .teachforamerica.org/about-us/our-mission/our-impact. I owe this observation to my students Olivia Petrocco, Jesse McGleughlin, and Olaitan Oladipo.

32. M. E. J. Newman, *Networks: An Introduction* (Oxford: Oxford University Press, 2010), 109.

33. Mark Newman, Albert-László Barabási, and Duncan J. Watts, *The Structure and Dynamics of Networks* (Princeton: Princeton University Press, 2006), 4.

34. Watts, *Six Degrees*, 43.

35. Latour, *Reassembling the Social*, 131.

36. See Anna Munster, "Networked Diagrammatism: From Map and Model to the Internet as Mechanogram," in *An Aesthesia of Networks: Conjunctive Experience in Art and Technology* (Cambridge, MA: MIT Press, 2013), 19–43. Also see Terranova, *Network Culture*; and D. Liben-Nowell and J. Kleinberg, "Tracing Information Flow on a Global Scale Using Internet Chain-letter Data," *Proceedings of the National Academy of Sciences of the United States of America* 105 (12) (2008): 4633–4638.

37. See Donna J. Haraway, "A Cyborg Manifesto: Science, Technology, and Socialist-Feminism in the Late Twentieth Century," in her *Simians, Cyborgs, and Women: The Reinvention of Nature* (New York: Routledge, 1991).

38. Jameson, *Postmodernism*, 38.

39. Claude E. Shannon and Warren Weaver, *The Mathematical Theory of Communication* (Urbana: University of Illinois Press, 1949).

40. See Sigmund Freud's "A Note upon the 'Mystic Writing Pad,'" in *General Psychological Theory: Papers on Metapsychology*, ed. Philip Rieff (New York: Touchstone, 1963). Freud writes that memory traces are stored in the pad. Memory and storage are not the same things. "Storage" usually refers to something material or substantial, as well as to its physical location: a store is both what and where it is stored. According to the *OED*, "to store" is to furnish, to build stock. Storage or stocks always look toward the future. Memory is not static, but rather is an active process: a memory must be held—constantly recalled—in order to keep it from fading. "Memory" stems from the same root as "martyr" and the ancient Greek term for "baneful" or "fastidious." Memory is an act of commemoration—the process of recollecting or remembering. With von Neumann architecture, computers have become nervous human machines, composed of amputated, simplified, and idealized neurons that contain within themselves a hierarchy of memory organs. The highest (or lowest) of these organs was the outside world: the input or output. Through this architecture, the world became "dead storage." Following this logic, to remain, things had to become volatile. Of course, there is also a forensics of reading deleted material. For more on this forensics, see Matthew Kirschenbaum's *Mechanisms: New Media and the Forensic Imagination* (Cambridge, MA: MIT Press, 2008).

41. For more on this idea, see Wendy Hui Kyong Chun, *Programmed Visions: Software and Memory* (Cambridge, MA: MIT Press, 2011).

42. See Gilles Deleuze, "Repetition for Itself," in *Difference and Repetition*, trans. Paul Patton (New York: Columbia University Press, 1994).

43. Gilles Deleuze, *Empiricism and Subjectivity: An Essay on Hume's Human Nature*, trans. Constantin V. Boundas (New York: Columbia University Press, 2001), 67.

44. Warren Weaver, "Recent Contributions to the Mathematical Theory of Communication," in Shannon and Weaver, *The Mathematical Theory of Communication*, 3–28.

45. For more on this, see "Part I: The Empirical Study of Networks," in Newman, *Networks: An Introduction*, 15–196.

46. Shannon and Weaver, *The Mathematical Theory of Communication*, 3–28.

47. See Jonah Lehrer, "Trials and Errors: Why Science Is Failing Us," *Wired*, December 16, 2011, n.p., http://www.wired.com/2011/12/ff_causation/; and Robert Mark et al., "Big Data, Systemic Risk and the U.S. Intelligence Community," *Enterprise Risk Management Symposium* (Chicago, April 22–24, 2013), n.p., accessed October 18, 2015, https://www.yumpu.com/en/document/view/15445772/2013-chicago-erm-2f-mark.

48. Deleuze, *Empiricism and Subjectivity*, 65.

49. David Hume, cited in ibid., 67.

50. Ibid.

51. Ibid., 68.

52. Ibid.

53. Hume, cited in ibid., 69.

54. Ibid.

55. Ibid., 69, 71.

56. Hume, cited in ibid., 71.

57. Chris Anderson, "The End of Theory: Will the Data Deluge Make the Scientific Method Obsolete?," *Edge*, June 30, 2009, n.p., http://edge.org/3rd_culture/anderson08/anderson08_index.html.

58. Ibid.

59. See Lisa Gitelman, ed., *"Raw Data" Is an Oxymoron* (Cambridge, MA: MIT Press, 2013).

60. See Viktor Mayer-Schönberger and Kenneth Cukier, *Big Data: A Revolution That Will Transform How We Live, Work, and Think* (New York: Houghton Mifflin Harcourt, 2013).

61. Ibid., 56–58.

62. See Charles Duhigg, "How Companies Learn Your Secrets," *New York Times*, February 16, 2012, http://www.nytimes.com/2012/02/19/magazine/shopping-habits.html?pagewanted=all.

63. Mayer-Schönberger and Cukier, *Big Data*, 64. By relying on correlations, Big Data can predict the future because it gives up on causality as a limiting factor: "there is nothing causal between car ownership and taking antibiotics as directed; the link between them is pure correlation. But findings as such were enough to inspire FICO's chief executive to boast in 2011, 'We know what you're going to do tomorrow'" (ibid., 56). Mayer-Schönberger and Cukier offer as evidence the already canonical examples of Big Data's success: in addition to Target's exposing a girl's pregnancy to her father before she did, they cite statistical methods which produce better translations than grammatical ones and Amazon's automatic recommendation system, which resulted in more purchases than its "editor choices."

64. As quoted by Antoinette Rouvroy in "Technology, Virtuality and Utopia: Governmentality in an Age of Autonomic Computing," in *The Philosophy of Law Meets the Philosophy of Technology: Autonomic Computing and Transformations of Human Agency*, ed. Mireille Hildebrandt and Antoinette Rouvroy (Milton Park, UK: Routledge, 2011), 128.

65. Ibid., 127.

66. Ibid.

67. See Philip Agre, "Surveillance and Capture: Two Models of Privacy," *Information Society: An International Journal* 10, no. 2 (1994): 101–127.

68. Ibid., 106.

69. Ibid.

70. Ibid.

71. This happens through the following five-stage cycle, which I explain here in terms of the steps involved in UPS's business model:

> 1. Analysis: in this stage, someone analyzed what is involved in an activity and breaks it down into steps (initial registration of the package; movement between centers, delivery, confirmation of delivery, etc.).
>
> 2. Articulation: someone then articulates how these acts (loading of van, movement between centers, delivery areas) are to be handled—this articulation, however, is usually framed more neutrally as a "discovery."
>
> 3. Imposition: this articulation is then imposed as the way things should be done.
>
> 4. Instrumentation: this process is then instrumentalized through various technologies (UPS puts in place a barcode to parse the action in nearly real time).
>
> 5. Elaboration: this stage deals with statistical analysis and further optimization (UPS creates activity records that are then stored, inspected, merged with other records).

72. Agre, "Surveillance and Capture," 112.

73. See also Wendy Brown, *Undoing the Demos: Neoliberalism's Stealth Revolution* (Cambridge, MA: MIT Press, 2015).

74. More fully, Agre argues that capture is not surveillance because:

> 1. Capture employs linguistic metaphors for human activity rather than visual ones
>
> 2. It assumes that the linguistic "parsing" of human activity actively intervenes into the process; it deliberately reorganizes those activities; whereas surveillance models assume that watching is nondisruptive and surreptitious. (Michel Foucault would disagree on this point with Agre: the possibility of being watched was meant to actively shape the one under surveillance, so that the inmate becomes his own captor.)
>
> 3. Capture uses structural metaphors, such as the catalog rather than territorial metaphors.

4. It is aligned with decentralized and heterogeneous organizations, rather than the state. Agre sees capture as a way of discussing organizations and effects without bringing in the question of state terror: the question of the state is both too big and too little.

5. Capture's driving aim is not political but philosophical: the point is to understand ontology, the essence of an activity. (See Agre, "Surveillance and Capture," 105–106.)

75. See Jacques Rancière, "Does Democracy Mean Something?," in *Dissensus: On Politics and Aesthetics*, trans. Steve Corcoran (New York: Continuum, 2010).

76. Elizabeth Bernstein, "Brokered Subjects and 'Illicit Networks,'" invited lecture, Conference on Habits of Living, Brown University, Providence, RI, March 2013.

Crisis + Habit = Update

1. Robert McFadden, "Vast Anti-Bush Rally Greets Republicans in New York," *New York Times*, August 30, 2004, n.p., http://www.nytimes.com/2004/08/30/politics/campaign/30protest.html.

2. Robert McFadden, "City Is Rebuffed on the Release of '04 Records," *New York Times*, August 7, 2007, http://www.nytimes.com/2007/08/07/nyregion/07police.html?ref=nationalspecial3.

3. Michelle Goldberg, "New York Lockdown," *Guardian*, August 11, 2004, http://www.theguardian.com/politics/2004/aug/11/education.

4. Tad Hirsch and John Henry, "TXTmob: Text Messaging for Protest Swarms," *Proceedings CHI EA '05 CHI '05 Extended Abstracts on Human Factors in Computing Systems* (New York: ACM Publications, 2005), 1455.

5. Ibid.

6. Ibid., 1456–1458.

7. Peter Rothberg, "Protest Pit Is a Dark, Shadowy Place," *Nation*, July 27, 2004, http://www.thenation.com/blog/protest-pit-dark-shadowy-place#.

8. Hirsch and Henry, "TXTmob," 1457.

9. Ibid., 1457–1458.

10. Ibid., 1457.

11. Ibid., 1458.

12. Colin Moynihan, "City Subpoenas Creator of Text Messaging Code," *New York Times*, March 30, 2008, http://www.nytimes.com/2008/03/30/nyregion/30text.html?ex=1364616000&en=aba124f5b62dae3c&ei=5124&partner=permalink&exprod=permalink.

13. maha, "Protesting 101," *The Mahablog*, April 12, 2006, http://www.mahablog.com/2006/04/12/protesting-101/.

14. Jet Jaguar, "Bizarre Photos and Quotes from Protestors at RNC Convention (Vanity)," *FreeRepublic.com*, August 29, 2004 (last modified September 9, 2004), http://www.freerepublic.com/focus/news/1202228/posts.

15. Haley Draznin, "New York to Pay $17.9 Million to 2004 Republican Convention Protestors," *CNN.com*, January 16, 2014, http://edition.cnn.com/2014/01/15/politics/new-york-republican-convention-settlement/.

16. Author's conversation with Tad Hirsch, February 13, 2015.

17. Again, "twitter" was originally a derogatory term: girls and birds tweeted.

18. Axel Bruns, "Crisis Communication, Social Media, and the Environment," in *The Media and Communications in Australia*, ed. Stuart Cunningham and Sue Turnbull (Sydney, Australia: Allen & Unwin, 2014), 351–355; Axel Bruns, "At Times of Crisis, Twitter Shines Brightest," in *Continuity* (Brisbane, Australia: Business Continuity Institute, 2012), 16–17; Y. Jin et al., "Examining the Role of Social Media in Effective Crisis Management," *Communications Research* 41, no. 1 (2014): 74–94.

19. Jin, "Examining the Role of Social Media," 76.

20. Anil Dash, *Twitter* (March 17, 2014), https://twitter.com/anildash/status/445761831597273088. For more on the conspiracy plots, see Barbara Tasch, "Former Malaysian Leader Accuses CIA of Cover-Up in Missing Jet," *Time.com*, May 19, 2014, http://time.com/104480/malaysia-airliens-flight-370-mahathir-mohamad/; and Sam Frizell, "The Missing Malaysian Plane: Five Conspiracy Theories," *Time.com*, March 11, 2014, http://time.com/20351/missing-plane-flight-mh370-5-conspiracy-theories/.

21. Jin, "Examining the Role of Social Media," 75.

Chapter 2

1. Milton Friedman, *Capitalism and Freedom* (Chicago: University of Chicago Press, 1982), ix.

2. Lauren Berlant, *Cruel Optimism* (Durham: Duke University Press, 2011), 3.

3. See Giorgio Agamben, *State of Exception*, trans. Kevin Attell (Chicago: University of Chicago Press, 2005).

4. As Barbara Johnson notes in her explanation of Jacques Derrida's critique of logocentrism, *logos* is the "image of perfectly self-present meaning ..., the underlying ideal of Western culture. Derrida has termed this belief in the self-presentation of meaning 'Logocentrism,' for the Greek word *Logos* (meaning speech, logic, reason, the Word of God)." See Barbara Johnson, "Translator's Introduction," in Jacques

Derrida, *Dissemination*, trans. Barbara Johnson (Chicago: University of Chicago Press, 1981), ix.

5. See Clare Carlisle, "Creatures of Habit: The Problem and the Practice of Liberation," *Continental Philosophy Review* 38, nos. 1–2 (2006): 25.

6. "Crisis," *OED*.

7. See Bill Gates, "Friction-Free Capitalism," in *The Road Ahead* (New York: Viking Press, 1995).

8. John Perry Barlow, "A Declaration of the Independence of Cyberspace," *EFF.org*, February 8, 1996, https://projects.eff.org/~barlow/Declaration-Final.html.

9. Ibid.

10. See Wendy Hui Kyong Chun, *Control and Freedom: Power and Paranoia in the Age of Fiber Optics* (Cambridge, MA: MIT Press, 2006).

11. See Al Gore, "Information Superhighways Speech," International Telecommunications Union, March 21, 1994, *AMDOCS: Documents for the Study of American History*, http://vlib.iue.it/history/internet/algorespeech.html.

12. Immanuel Kant, "An Answer to the Question: What Is Enlightenment?," in *What Is Enlightenment? Eighteenth-Century Answers and Twentieth-Century Questions*, ed. James Schmidt (Berkeley: University of California Press, 1996), 58–64.

13. For more on enlightenment as a stance of how not to be governed "like that," see Michel Foucault, "What Is Critique?," in Schmidt, *What Is Enlightenment?*, 382–398.

14. For examples, see George Landow, *Hypertext: The Convergence of Contemporary Critical Theory and Technology* (Baltimore: Johns Hopkins University Press, 1992); Sherry Turkle, *Life on the Screen: Identity in the Age of the Internet* (New York: Simon and Schuster, 1995).

15. During a congressional debate over the Communications Decency Act, Senator Daniel R. Coats argued: "perfunctory onscreen warnings which inform minors they are on their honor not to look at this [are] like taking a porn shop and putting it in the bedroom of your children and then saying 'Do not look.'" As quoted in Department of Justice Brief (*Reno v. ACLU*), January 21, 1997. http://www.ciec.org/SC_appeal/970121_DOJ_brief.html.

16. "Godwin's Law," *Wikipedia.com*, last modified October 11, 2015, http://en.wikipedia.org/wiki/Godwin's_law.

17. For more on this, see Chun, *Control and Freedom*.

18. See Kang Hyun-Kyong, "Cell Phones Create Youth Nationalism," *Korea Times Online*, May 12, 2008, http://koreatimes.co.kr/www/news/special/2008/06/180 _24035.html.

19. See McKenzie Wark, "The Weird Global Media Event and the Tactical Intellectual," in *New Media, Old Media: A History and Theory Reader*, ed. Wendy Hui Kyong Chun and Thomas Keenan (New York: Routledge, 2006), and Geert Lovink, "Enemy of Nostalgia: Victim of the Present, Critic of the Future," *PAJ: A Journal of Performance and Art* 24, no. 1 (2002): 5–15. Lovink contends, "because of the speed of events, there is a real danger that an online phenomenon will already have disappeared before a critical discourse reflecting on it has had the time to mature and establish itself as institutionally recognized knowledge," in *My First Recession: Critical Internet Culture in Transition* (Amsterdam: Institute of Network Cultures, 2011), 8.

20. Wark, "The Weird Global Media Event," 265.

21. For more on this, see Wendy Hui Kyong Chun, "The Enduring Ephemeral, or the Future Is a Memory," *Critical Inquiry* 35, no. 1 (2008): 151–152.

22. See Ursula Frohne, "Screen Tests: Media, Narcissism, Theatricality, and the Internalized Observer," in *CTRL [Space]: Rhetorics of Surveillance from Bentham to Big Brother*, ed. Thomas Levin, Ursula Frohne, and Peter Weibel (Cambridge, MA: MIT Press, 2002), 252.

23. Jane Feuer, "The Concept of Live Television: Ontology as Ideology," in *Regarding Television: Critical Approaches*, ed. E. A. Kaplan (Washington: University Press of America, 1983), 12–22.

24. Mary Ann Doane, "Information, Crisis, Catastrophe," in Chun and Keenan, *New Media, Old Media*, 262.

25. Ibid.

26. Televisual catastrophe is thus "characterized by everything which it is said not to be—it is expected, predictable, its presence crucial to television's operation. ... [C]atastrophe functions as both the exception and the norm of a television practice which continually holds out to its spectator the lure of a referentiality perpetually deferred." Ibid., 262.

27. See Thomas Levin, "Rhetoric of the Temporal Index: Surveillant Narration and the Cinema of 'Real Time,'" in Levin, Frohne, and Weibel, *CTRL [Space]*, 578–593.

28. See Tara McPherson, "Reload: Liveness, Mobility and the Web," in *The Visual Culture Reader*, ed. Nicholas Mirzoeff, 2nd ed. (New York: Routledge, 2004), 462.

29. See McPherson, "Reload: Liveness," and Alexander Galloway, *Protocol: How Control Exists after Decentralization* (Cambridge, MA: MIT Press, 2004), 54–79.

30. See Brett M. Christensen, "Osama Bin Laden Virus Emails," July 2, 2005; last updated May 10, 2011, http://www.hoax-slayer.com/bin-laden-captured.html.

31. See David Liben-Nowell and Jon Kleinberg, "Tracing Information Flow on a Global Scale Using Internet Chain-Letter Data," *Proceedings of the National Academy of Sciences* 105, no. 12 (2008): 4633–4638.

32. See Berlant, *Cruel Optimism*, 4.

33. This notion of crisis ordinary engages and revises the classic Marxist notion that capitalism creates crises, which will lead to its undoing. Crisis become ordinary becomes a chronic condition in the era of neoliberalism, and this chronicity leads to a steady dulling of the future.

34. *OED*, s.v. "real time."

35. Or, as Slavoj Žižek puts it, the danger "is thus that the predominant narrative of the meltdown will be the one which, instead of awakening us from a dream, will enable us to *continue dreaming*." See Žižek, *First as Tragedy, Then as Farce* (London: Verso, 2009), 20.

36. Jacques Derrida, "Force of Law: The Mystical Foundation of Authority," in *Acts of Religion*, ed. Gil Anidjar (New York: Routledge, 2002), 252.

37. See Agamben, *State of Exception*, 7.

38. Ibid., 40.

39. Ibid., 36. According to Agamben, "The state of exception is an anomic space in which what is at stake is a force of law without law (which should therefore be written: force-of-~~law~~). Such a 'force-of-~~law~~,' in which potentiality and act are radically separated, is certainly something like a mystical element, or rather a *fictio* by means of which law seeks to annex anomie itself" (ibid., 39).

40. For more twists—such as memory and program—see "Order from Order, or Life According to Software" and "Always Already There, or Software as Memory" in Wendy Hui Kyong Chun, *Programmed Visions: Software and Memory* (Cambridge, MA: MIT Press, 2011).

41. See Wendy Wood, Jennifer S. Labrecque, Pei-Ying Lin, and Dennis Rünger, "Habits in Dual-Process Models," in *Dual-Process Theories of the Social Mind*, ed. Jeffrey W. Sherman, Bertram Gawronski, and Yaacov Trope (New York: Guilford Press, 2014), 372.

42. Ibid., 374.

43. Ibid.

44. See Ann Graybiel, "The Basal Ganglia and Chunking of Action Repertoires," *Neurobiology of Learning and Memory* 70, nos. 1–2 (1998): 120.

45. Ibid., 127.

46. Wood et al., "Habits in Dual-Process Models," 374–375.

47. See Carlisle, "Creatures of Habit," 26.

48. Chris Loesch and Keith Payne, "The Situated Inference Model of Priming: How a Single Prime Can Alter Perception, Goals, and Behavior," talk given at the Society for Personality and Social Psychology Annual Meeting, Las Vegas, NV (January 2010), quoted in Wendy Wood and David T. Neal, "A New Look at Habits and the Habit-Goal Interface," *Psychological Review* 114, no. 4 (2007): 843–863.

49. Graybiel, "The Basal Ganglia and Chunking of Action Repertoires," 131–132.

50. For more on this, see "On Sourcery and Source Codes," in Chun, *Programmed Visions*, 19–54. This privileging of code is evident in common sense and theoretical understandings of programming, from claims made by free software advocates that free source code is freedom, to those made by new media theorists that new media studies is, or should be, software studies. Programmers, computer scientists, and critical theorists have all reduced software—once evocatively described by historian Michael Mahoney as "elusively intangible ... the behavior of the machines when running" and described by theorist Adrian Mackenzie as a "neighborhood of relations"—to a recipe, a set of instructions, substituting space/text for time/process. See Michael Mahoney, "The History of Computing in the History of Technology," *IEEE Annals of the History of Computing* 10, no. 2 (1988): 121; and Adrian Mackenzie, *Cutting Code: Software and Sociality* (New York: Peter Lang, 2006), 169.

51. Several new media theorists have theorized code as essentially and rigorously "executable." Alexander Galloway, for instance, has powerfully argued, "code draws a line between what is material and what is active, in essence saying that writing (hardware) cannot *do* anything, but must be transformed into code (software) to be effective. ... Code is a language, but a very special kind of language. *Code is the only language that is executable*. ... [C]ode is the first language that actually does what it says." See Galloway, *Protocol*, 165–166, emphasis in original. Given that the adjective "executable" applies to anything that "can be executed, performed, or carried out" (the first example of "executable" given by the *OED* is from 1796), this is a strange statement.

52. See Jacques Derrida's analysis of *Phaedrus* in his essay "Plato's Pharmacy," in *Dissemination*, 134.

53. Not surprisingly, this notion of source code as source coincides with the introduction of alphanumeric languages. With them, human-written, nonexecutable code becomes source code, and the compiled code becomes the object code. Source code, therefore, is symptomatic of human language's tendency to attribute a sovereign source to an action, a subject to a verb. Thus Galloway's statement, "to see code as subjectively performative or enunciative is to anthropomorphize it, to project it onto the rubric of psychology, rather than to understand it through its own logic of 'calculation' or 'command'" overlooks the fact that to use higher-level alphanumeric languages is already to anthropomorphize the machine and to reduce all machinic actions to the commands that supposedly drive them. See Alexander Galloway, *The Interface Effect* (Cambridge: Polity, 2012), 71.

54. See Lawrence Lessig, *Code and Other Laws of Cyberspace* (New York: Basic Books, 1999).

55. *OED*, s.v. "code."

56. Derrida, "Force of Law," 279.

57. Ibid., 278.

58. See Judith Butler, *Excitable Speech: A Politics of the Performative* (New York: Routledge, 1997), 48–49.

59. Ibid., 78.

60. Fred Brooks, while responding to the disaster that was OS/360, also emphasized the magical powers of programming. Describing the joys of the craft, Brooks writes:

> Why is programming fun? What delights may its practitioner expect as his reward?
> First is the sheer joy of making things. ...
> Second is the pleasure of making things that are useful to other people. ...
> Third is the fascination of fashioning complex puzzle-like objects of interlocking moving parts and watching them work in subtle cycles, playing out the consequences of principles built in from the beginning. ...
> Fourth is the joy of always learning, which springs from the nonrepeating nature of the task. ...
> Finally there is the delight of working in such a tractable medium. The programmer, like the poet, works only slightly removed from thought-stuff. He builds his castles in the air, from air, creating by exertion of the imagination. ... Yet the program construct, unlike the poet's words, is real in the sense that it moves and works, producing visible outputs separate from the construct itself. It prints results, draws pictures, produces sounds, moves arms. The magic of myth and legend has come true in our time. One types the correct incantation on a keyboard, and a display screen comes to life, showing things that never were nor could be.

See Fred Brooks, *The Mythical Man-Month: Essays on Software Engineering* (Reading, MA: Addison-Wesley Professional, 1995), 7–8.

61. See Joseph Weizenbaum, *Computer Power and Human Reason: From Judgment to Calculation* (San Francisco: W. H. Freeman, 1976), 115.

62. Derrida, "Force of Law," 281.

63. Agamben, *State of Exception*, 82.

64. Ibid., 76.

65. Ibid., 86.

66. Derrida, "Force of Law," 293.

67. Butler, *Excitable Speech*, 39.

68. See Wendy Wood, Jeffrey M. Quinn, and Deborah A. Kashy, "Habits in Everyday Life: Thought, Emotion, Action," *Journal of Personality and Social Psychology* 83, no. 6 (2002): 1281–1297; and Charles Duhigg, *The Power of Habit: Why We Do What We Do in Life and Business* (New York: Random House, 2012).

69. See Ian Burkitt, "Technologies of the Self: Habitus and Capacities," *Journal for the Theory of Social Behavior* 32, no. 2 (2002): 220–237.

70. Wood et al., "Habits in Dual-Process Models," 375; Elizabeth Grosz, in her reading of Henri Bergson, similarly argues that habits are "memories that are activated unconsciously and without effort as preparatory for action"; see Grosz, "Habit Today: Ravaisson, Bergson, Deleuze and Us," *Body and Society* 19, nos. 2–3 (2013): 228.

71. See Naomi Klein, *Shock Doctrine: The Rise of Disaster Capitalism* (New York: Picador, 2007).

72. For more on this, see Chun, *Programmed Visions*.

73. See John von Neumann, "The First Draft Report on the EDVAC," Contract No. W-670-ORD-4926 between the USORD and the University of Pennsylvania, June 30, 1945, http://www.virtualtravelog.net/wp/wp-content/media/2003-08-TheFirstDraft.pdf.

74. Eric Kandel, *In Search of Memory: The Emergence of a New Science of Mind* (New York: Norton, 2006).

75. K. B. McDermott, "Implicit Memory," in *Encyclopedia of Psychology*, ed. Alan E. Kazdin (Washington, DC: American Psychological Association; Oxford: Oxford University Press, 2000), 4: 231.

76. Kandel, *In Search of Memory*, 132.

77. Ibid., 133.

78. This happens due to the transmission of seratonin, which leads to the production of cyclic AMP, which in turn frees the catalytic unit protein kinase A, which then enhances the release of the neurotransmitter.

79. This is because repeated stimulation causes the kinases to move into the nucleus, leading to the expression of the CREB protein, which then generates new synapses.

80. Kandel, *In Search of Memory*, 215. As Kandel puts it, if we remember anything of this book, it is because our brain has changed.

81. Ibid., 313.

82. This separation of voluntary from involuntary memory/emotions maps nicely onto theoretical ruminations on the relationship between affect and feelings, or involuntary and voluntary emotions. Affects are involuntary and caused by factors other than the self, such as circulating signals. This fact arguably reveals why crises are so privileged "Now": they are moments when both are active; moments when new pathways can be built.

83. See Adnia Roskies, "Are Neuroimages Like Photographs of the Brain?," *Philosophy of Science* 74, no. 5 (2007): 860–872; and Gabriella Coleman, "The Politics of

Rationality: Psychiatric Survivors' Challenge to Psychiatry," in *Tactical Biopolitics: Art, Activism, and Technoscience*, ed. Beatriz da Costa and Kavita Philip (Cambridge, MA: MIT Press, 2008).

84. Kandel, *In Search of Memory*, 260.

85. See Endel Tulving, *Elements of Episodic Memory* (Oxford: Oxford University Press, 1983), 7.

86. See Henry Roediger, "Reconsidering Implicit Memory," in *Rethinking Implicit Memory*, ed. Jeffrey Bowers and Chad Marsolek (Oxford: Oxford University Press, 2003), 3–18.

87. See Wood, Quinn, and Kashy, "Habits in Everyday Life."

88. Intriguingly, in describing explicit memory—in which we do not simply recall but rather experience an event once more—Kandel emphasizes creativity. The brain, he asserts, only stores a core memory. Upon recall, though, this core memory is "elaborated upon and reconstructed, with subtractions, additions, elaborations, and distortions" (*In Search of Memory*, 281). That which persists—that which is recalled and that traumatizes—we experience each time as though it were the first time, as though it just happened. Memories are persistent and will not leave us alone; they are not simply missives from the past, but repetitions that are created each time.

89. Kandel, *In Search of Memory*, 10.

90. Ibid., 160.

91. Félix Ravaisson, *Of Habit*, trans. C. Carlisle and M. Sinclair (London: Continuum, 2008).

92. Grosz, "Habit Today," 219.

93. Ibid., 233.

94. Wolfgang Ernst thus argues that new media are time-based. See Wolfgang Ernst, "Dis/continuities: Does the Archive Become Metaphorical in Multi-Media Space?," in Chun and Keenan, *New Media, Old Media*.

95. Thomas Keenan, *Fables of Responsibility: Aberrations and Predicaments in Ethics and Politics* (Stanford: Stanford University Press, 1997), 1.

96. Derrida, "Force of Law," 252.

97. See Jacques Derrida, "Cogito and the History of Madness," in *Writing and Difference*, trans. Alan Bass (Chicago: University of Chicago Press, 1978), 31.

98. This deferral of decision stemming from a belief in information as decision catches us in a deluge of minor-seeming decisions that defer our engagement with crisis, or rather, render everything and thus nothing a crisis.

99. See Antoinette Rouvroy, "Technology, Virtuality and Utopia: Governmentality in an Age of Autonomic Computing," in *The Philosophy of Law Meets the Philosophy*

of Technology: Autonomic Computing and Transformations of Human Agency, ed. Mireille Hildebrandt and Antoinette Rouvroy (Milton Park, UK: Routledge, 2011).

100. For more on this and models as "hypo-real," see Wendy Hui Kyong Chun, "Hypo-Real Models, or Global Climate Change: A Challenge for the Humanities," *Critical Inquiry* 41, no. 3 (Spring 2015): 675–703.

101. Derrida, "Force of Law," 253.

Part II

1. Glenn Greenwald, "NSA Collecting Phone Records of Millions of Verizon Customers Daily," *Guardian*, June 6, 2013, http://www.theguardian.com/world/2013/jun/06/nsa-phone-records-verizon-court-order.

2. "AT&T Whistle-Blower's Evidence," *Wired.com*, May 17, 2006, http://archive.wired.com/science/discoveries/news/2006/05/70908.

3. *Smith v. Maryland*, 422 U.S. 735 (1979), accessed October 20, 2015, http://caselaw.findlaw.com/us-supreme-court/442/735.html.

4. "Access to certain business records for foreign intelligence and international terrorism investigations," U.S. Code Title 50, Chapter 36, Subchapter IV, § 1861, *Legal Information Institute*, Cornell University, accessed October 20, 2015, https://www.law.cornell.edu/uscode/text/50/1861.

5. "Snowden Leaks: Google 'Outraged' at Alleged NSA Hacking," *BBC News*, October 31, 2013, http://www.bbc.com/news/world-us-canada-24751821.

6. D. A. Miller, *The Novel and the Police* (Berkeley: University of California Press, 1988), 208.

7. Ibid.

8. Thomas Keenan, "Windows: Of Vulnerability," in *The Phantom Public Sphere*, ed. Bruce Robbins (Minneapolis: University of Minnesota Press, 1997), 132.

9. Ibid.

10. The public "is the experience, if we can call it that, of the interruption or the intrusion of all that is radically irreducible to the order of the individual human subject, the unavoidable entrance of alterity into the everyday life of the 'one' who would be human. … Publicity tears us from our selves, exposes us to and involves us with others, denies us the security of that window behind which we might install ourselves to gaze." Ibid., 133–134. Crucially, in addition to breaching the public and the private, this "tearing" also makes possible the public and the private, intersubjectivity and interiority. Publicity does not contaminate or open up "an otherwise sealed interiority. Rather, what we call interiority is itself the mark or the trace of this breach, of a violence that in turn makes possible the violence or the love we experience as

intersubjectivity. We would have no relation to others, no terror and no peace, certainly no politics, without this (de)constitutive interruption." Ibid., 134.

11. Anne Friedberg, *The Virtual Window: From Alberti to Microsoft* (Cambridge, MA: MIT Press, 2006).

12. See Eve Sedgwick, *Epistemology of the Closet* (Berkeley: University of California Press, 1990).

13. "Once a person voluntarily places him- or herself in the public eye, that person cannot complain when he or she is given publicity that they have sought, even if the publicity is unfavorable." See Justin Lee, "The Reasonable Expectation of Invasion," *Science and Technology Law Quarterly* 1, no. 2 (2008), http://www.americanbar.org/newsletter/publications/scitech_e_merging_news_home/privacy.html#_edn15.

14. See Shilpa Phadke, Sameera Khan, and Shilpa Ranade, *Why Loiter? Women and Risk on Mumbai Streets* (New Delhi: Penguin Books, 2011), a text discussed extensively in chapter 4.

The Friend of My Friend Is My Enemy (and Thus My Friend)

1. "Leaked Steubenville Big Red Rape Video," *YouTube.com*, https://www.youtube.com/watch?v=W1oahqCzwcY&bpctr=1400682132.

2. Juliet Macur and Nate Schweber, "Rape Case Unfolds on Web and Splits City," *New York Times*, December 16, 2012, http://www.nytimes.com/2012/12/17/sports/high-school-football-rape-case-unfolds-online-and-divides-steubenville-ohio.html?pagewanted=all&_r=0.

3. Alexandra Goddard, "I Am the Blogger Who Allegedly 'Complicated' the Steubenville Gang Rape Case—and I Wouldn't Change a Thing," *Xo Jane*, March 18, 2013, http://www.xojane.com/issues/steubenville-rape-verdict-alexandria-goddard.

4. Rebecca Shapiro, "Poppy Harlow, CNN Reporter, 'Outraged' over Steubenville Rape Coverage Criticism: Report," *Huffington Post*, March 20, 2013, http://www.huffingtonpost.com/2013/03/20/poppy-harlow-cnn-steubenville-rape-coverage-criticism_n_2914853.html.

5. David Kushner, "Anonymous vs. Steubenville," *Rolling Stone*, November 27, 2013, http://www.rollingstone.com/culture/news/anonymous-vs-steubenville-20131127.

6. Bianca Bosker, "Facebook's Randi Zuckerberg: Anonymity Online 'Has to Go Away,'" *Huffington Post*, July 27, 2011, http://www.huffingtonpost.com/2011/07/27/randi-zuckerberg-anonymity-online_n_910892.html.

7. Michel Foucault, *Discipline and Punish: The Birth of the Prison*, trans. Alan Sheridan (New York: Vintage Books, 1995), 203.

8. Ibid., 201.

9. Ibid., 202.

10. Ibid., 219.

11. Ibid., 211–212.

12. Gabriella Coleman, "Our Weirdness Is Free: The Logic of Anonymous—Online Army, Agent of Chaos, and Seeker of Justice," *Triple Canopy*, January 2012, accessed August 30, 2014, http://canopycanopycanopy.com/issues/15/contents/our_weirdness_is_free.

13. See Danielle Allen, "Anonymous: On Silence and the Public Sphere," in *Speech and Silence in American Law*, ed. Austin Sarat (New York: Cambridge University Press, 2010), 124.

14. See Laura Poitras's *Citizenfour* (Praxis Films, 2014).

15. See Jürgen Habermas, *The Structural Transformation of the Public Sphere: An Inquiry into a Category of Bourgeois Society*, trans. Thomas Burger and Frederick Lawrence (Cambridge, MA: MIT Press, 1991).

16. Eden Osucha, "The Whiteness of Privacy: Race, Media, Law," *Camera Obscura* 24, no. 1 (2009): 67–107.

Chapter 3

1. See Oskar Negt and Alexander Kluge, *Public Sphere and Experience: Toward an Analysis of the Bourgeois and Proletarian Public Sphere*, trans. Peter Labanyo and Assenka Oksiloff (Minneapolis: University of Minnesota Press, 1993), 54.

2. See Elizabeth Povinelli, *The Empire of Love: Toward a Theory of Intimacy, Genealogy, and Carnality* (Durham: Duke University Press, 2006), 190.

3. Margaret Thatcher in Douglas Keay, "Margaret Thatcher Interview," *Woman's Own* (1987), http://www.margaretthatcher.org/speeches/displaydocument.asp?docid=106689.

4. Jacques Derrida in *The Politics of Friendship* argues, "friendship ... is first accessible on the side of its subject, who thinks and lives it, not on the side of its object, who can be loved or loveable without in any way being assigned to a sentiment of which, precisely, he remains the object. ... This incommensurability between the lover and the beloved will now unceasingly exceed all measurement and all moderation—that is, it will exceed the very principle of a calculation." See Derrida, *The Politics of Friendship* (New York: Verso, 1997), 10.

5. See John Perry Barlow, "A Declaration of the Independence of Cyberspace," *EFF.org*, February 8, 1996, https://projects.eff.org/~barlow/Declaration-Final.html.

6. William Gibson, *Neuromancer* (New York: Ace Books, 1984), 52.

7. Ibid., 6.

8. For more on this, see "Why Cyberspace?," in Wendy Hui Kyong Chun, *Control and Freedom: Power and Paranoia in the Age of Fiber Optics* (Cambridge, MA: MIT Press, 2006), 37–76.

9. See Chun, *Control and Freedom*.

10. For more on the relationship between semiprivate spaces and critical thinking, see Ellen Rooney, "A Semiprivate Room," *Differences: A Journal of Feminist Cultural Studies* 13, no. 1 (2002): 128–156.

11. See Geert Lovink, *My First Recession: Critical Internet Culture in Transition* (Amsterdam: Institute of Network Cultures, 2011).

12. See Chun, "Screening Pornography," in *Control and Freedom*, 77–128.

13. Ibid.

14. Abrams v. United States, 250 U.S. 616, 630 (1919) (Holmes, J., dissenting).

15. Reno v. ACLU, Supreme Court, Department of Justice brief, No. 96-511, filed January 21, 1997

16. See Bianca Bosker, "Facebook's Randi Zuckerberg: Anonymity Online 'Has to Go Away,'" *Huffington Post*, July 27, 2011, http://www.huffingtonpost.com/2011/07/27/randi-zuckerberg-anonymity-online_n_910892.html; and Terrell Ward Bynum, "Anonymity on the Internet and Ethical Accountability," *The Research Center on Computing and Society* (1997), posted September 5, 2010, http://rccs.southernct.edu/on-the-emerging-global-information-ethics/. Importantly, this was not the only solution offered to foster critical public dialog: competing against this simple connection of transparency with responsibility were formalized "reputation systems," such as the one developed by *Slashdot.org*, which were based on pseudonymic usage. These systems evolved through long-term use and communal evaluation, two features that would prove essential to any functioning online space, whether pseudonymic or transparent.

17. See Bosker, "Facebook's Randi Zuckerberg: Anonymity Online 'Has to Go Away'"; and Bianca Bosker, "Eric Schmidt on Privacy (VIDEO): Google CEO Says Anonymity Online Is 'Dangerous,'" *Huffington Post*, August 10, 2010, http://www.huffingtonpost.com/2010/08/10/eric-schmidt-privacy-stan_n_677224.html.

18. Helen Nissenbaum, "Securing Trust Online: Wisdom or Oxymoron?," *Boston University Law Review* 81, no. 3 (June 2001): 655.

19. Ibid., 662.

20. To cite two of the more canonical examples: after Phoebe Prince, a fifteen-year-old transfer student at South Hadley High School in Massachusetts, hanged herself, comments such as "she deserved it" and "mission accomplished" appeared on her *Facebook.com* wall. In addition, several of her enemies created a "We Murdered

Phoebe Prince" page, featuring a digitally altered image of her, in which knives were shown piercing her eyes. The perpetrators were apparently four girls and one boy who were upset over Phoebe's romantic entanglements with their 'friends.' See Alyssa Giacobbe, "Who Failed Phoebe Prince?," *Boston Magazine* (June 2010), http://www.bostonmagazine.com/2010/05/phoebe-prince/. Later that year, Tyler Clementi jumped off the George Washington Bridge after his roommate Dharun Ravi and his hallmate Molly Wei, using Ravi's webcam, twice spied on Clementi's romantic activities with another man. After the first episode, Ravi had tweeted: "Roommate asked for the room till midnight. I went into molly's room and turned on my webcam. I saw him making out with a dude. Yay"; he also "dared" people with iChat to participate in the second viewing. See Lisa Foderaro, "Private Moment Made Public, Then a Fatal Jump," *New York Times*, September 29, 2010, http://www.nytimes.com/2010/09/30/nyregion/30suicide.html?_r=3&).

21. Janis Wolak, David Finkelhor, and Kimberly J. Mitchell, "Trends in Arrests for Child Pornography Production: The Third National Juvenile Online Victimization Study (NJOV-3)," *Crimes against Children Research Center*, Durham, NH (CV270), April 2012, http://www.unh.edu/ccrc/pdf/CV270_Child%20Porn%20Production%20Bulletin_4-13-12.pdf.

22. Amy Kimpel, "Using Laws Designed to Protect as a Weapon: Prosecuting Minors under Child Pornography Laws," *New York University Review of Law and Social Change* 34, no. 2 (2010): 299.

23. danah boyd, "Friendster and Publicly Articulated Social Networks," Conference on Human Factors and Computing Systems (CHI 2004) (ACM, Vienna, April 24–29, 2004), http://www.danah.org/papers/CHI2004Friendster.pdf.

24. Ibid.

25. Katherine Mieszkowski, "Faking out Friendster," *Salon*, August 14, 2003, http://www.salon.com/2003/08/14/fakesters/.

26. danah boyd and Jeffrey Heer, "Profiles as Conversation: Networked Identity Performance on Friendster," Proceedings of the Hawai'i International Conference on System Sciences (HICSS-39), Persistent Conversation Track Kauai, HI: IEEE Computer Society (January 4–7, 2006), http://vis.stanford.edu/files/2006-Friendster-HICSS.pdf.

27. Povinelli, *The Empire of Love*, 190.

28. Ibid.

29. See Derrida, *The Politics of Friendship*.

30. See Emmanuel Levinas, *Otherwise Than Being, or, Beyond Essence*, trans. Alphonso Lingis (Boston: Martinus Nijhoff, 1981), 127.

31. boyd, "Friendster and Publicly Articulated Social Networks."

32. For more on free labor, see Tiziana Terranova, "Free Labour," in *Network Culture: Politics for the Information Age* (London: Pluto Press, 2004), 73–97.

33. Michael Arrington, "85% of College Students Use FaceBook," *TechCrunch*, September 7, 2005, accessed August 30, 2014, http://techcrunch.com/2005/09/07/85-of-college-students-use-facebook/.

34. *MySpace.com*, of course, also engaged in a slippage between the private and public. Although profiles seem focused on the self (as the name implies), they depended on friend traffic and friend listings in order to generate interest.

35. *Google.com*'s 2012 decision to link its various databases that tracked user movement in seemingly separate spaces reveals the extent to which the Internet—even in its allegedly most open form—is increasingly a closed space. This logic of enclosure, however, was evident in *Google.com*'s initial form: a database. When one enters a term in *Google.com*, one does not search the Internet, but rather *Google.com*'s database, a simulacrum of the Internet.

36. As anthropologist Elizabeth Bernstein has argued in her work on the ramifications of changes to prostitution laws on the actual lives of sex workers, private spaces are often far more dangerous than public ones. Bernstein's notion of "bounded authenticity," of the transformation of prostitutes to "girlfriends for hire," also resonates with the transformation of friends outlined in this chapter. See Elizabeth Bernstein, *Temporarily Yours: Intimacy, Authenticity, and the Commerce of Sex* (Chicago: University of Chicago Press, 2007).

37. danah boyd et al., "Friendship," in *Hanging Out, Messing Around, Geeking Out: Kids Living and Learning with New Media*, ed. Mizuko Ito et al. (Cambridge, MA: MIT Press, 2008).

38. Ibid., 91.

39. Intriguingly, in terms of sex crimes involving juveniles, "the largest number of cases involving SNSs [social networking sites] was undercover operations in which investigators set up web pages and profiles in the course of portraying minors online." See Kimberly Mitchell et al., "Use of Social Networking Sites in Online Sex Crimes against Minors: An Examination of National Incidence and Means of Utilization," *Journal of Adolescent Health* 47, no. 2 (2010): 185. The profile of those caught in "sting" operations—which in 2003 constituted a larger group than those arrested for actually soliciting youth—intriguingly differed from other online molesters. They were "older and more middle class in income and employment compared with those who solicited actual youths. They were also somewhat less likely to have prior arrests for sexual offenses against minors or for nonsexual offenses or to have histories of violence or deviant sexual behavior." See Janis Wolak et al., "Online 'Predators' and Their Victims: Myths, Realities, and Implications for Prevention and Treatment," *American Psychologist* 63, no. 2 (February–March 2008): 119. Wolak et al. speculate that this difference might be to due to the fact that "the offenders

most likely to be fooled by undercover investigators lack suspicion about law enforcement because they have less criminal experience and higher social status. It could also be that some such individuals are less experienced or skilled and more naïve in their pursuit of youths and are thus more easily caught" (ibid., 119). SNSs, because they reek of a certain authenticity, seemed to be more effective at "catching" would-be naïve offenders.

40. Jessica Ringrose et al., *A Qualitative Study of Children, Young People and 'Sexting': A Report Prepared for the NSPCC* (London: NSPCC, 2012), 7.

41. Ibid., 13.

42. Ibid., 27.

43. Manuel Castells, "The New Economy: Informationalism, Globalization, Networking," in Castells, *The Rise of Network Society*, vol. 1 of *The Information Age* (New York: Blackwell, 2000), 77–162.

44. See, for example, Sanjeev Goyal, *Connections: An Introduction to the Network Economy* (Princeton: Princeton University Press, 2007).

45. Terranova, *Network Culture*, 90.

46. Walter Benjamin, "The Story Teller," in *Illuminations* (New York: Harcourt Brace Jovanovich, 1969), 90.

47. Yochai Benkler, *The Wealth of Networks: How Social Production Transforms Markets and Freedom* (New Haven: Yale University Press, 2006); Pierre Lévy, *Collective Intelligence: Mankind's Emerging World in Cyberspace* (New York: Plenum, 1997); and Paolo Virno, "General Intellect," in *Lessico postfordista* (Milan: Feltrinelli, 2001).

48. For more on this, see Mung Chiang, *Networked Life: 20 Questions and Answers* (Cambridge: Cambridge University Press, 2012).

49. See "The Netflix Prize Rules," *Netflixprize.com*, 2009, http://www.netflixprize.com/rules.

50. Allan Sekula, "The Body and the Archive: The Use and Classification of Portrait Photography by the Police and Social Scientists in the Late 19th and Early 20th Centuries," *October* 39 (1986): 3–64.

51. Casey Johnston, "Netflix Never Used Its $1 Million Algorithm due to Engineering Costs," *Wired*, April 16, 2012, http://www.wired.com/2012/04/netflix-prize-costs/.

52. Louis Althusser, *Lenin and Philosophy and Other Essays* (London: New Left Books, 1971), 162.

53. Jacques Lacan, "The Mirror Stage as Formative of the Function of the *I* as Revealed in Psychoanalytic Experience," in *Écrits: A Selection*, trans. Alan Sheridan (London: Tavistock, 1977), 1–7.

54. Althusser, *Lenin and Philosophy*, 174.

55. Ibid.

56. Richard Dienst, *Still Life in Real Time: Theory after Television* (Durham: Duke University Press, 1994). See in particular chapter 7, 128–143.

57. Ibid., 141.

58. Ibid.

59. Ibid., 141–142.

60. On the possibilities contained in the silence of the masses, see Jean Baudrillard, *In the Shadow of the Silent Majorities, or, The End of the Social, and Other Essays* (New York: Semiotext(e), 1983).

61. Ben Rooney, "Big Data's Big Problem: Little Talent," *Wall Street Journal*, April 29, 2012, http://www.wsj.com/articles/SB10001424052702304723304577365700368073674.

62. Adam, one of the organizers of a proposed London flash mob, commented: "Flash mobs anchor the online world into the real world—they are a manifestation of your 'cc' list." See Sandra Shmueli, "'Flash Mob' Craze Spreads," *CNN.com*, August 8, 2003, http://www.cnn.com/2003/TECH/internet/08/04/flash.mob/index.html?_s=PM:TECH.

63. Quoted in Leander Kahney, "E-mail Mobs Materialize All Over," *Wired*, May 7, 2003, http://archive.wired.com/culture/lifestyle/news/2003/07/59518.

64. See Jacques Rancière, *Disagreement: Politics and Philosophy*, trans. Julie Rose (Minneapolis: University of Minnesota Press, 1999).

65. Povinelli, *Empire of Love*, 35.

66. Quoted in Trent Mankelow, "Quotes from Webstock 2012," *Optimal Usability Blog*, February 23, 2012; accessed August 30, 2014, http://www.optimalusability.com/2012/02/quotes-from-webstock-2012/.

67. Jean-Luc Nancy, *The Experience of Freedom*, trans. Bridget McDonald (Stanford: Stanford University Press, 1993), xx.

I Never Remember; YOUs Never Forget

1. See "Maria Marroquin, Dream Activist Pennsylvania," *Youtube.Com* (April 5, 2011), https://www.youtube.com/watch?v=aFXTNk3YOnY. This video is arguably the most popular because an article by the Center for Immigrant Studies targeted her as an "illegal alien" working for Obamacare; see James R. Edwards, Jr., "Illegal Alien Heads Obamacare Navigator Program in NYC," *Center for Immigration Studies*, October 15, 2013, http://cis.org/edwards/illegal-alien-heads-obamacare-navigator-

program-nyc. Her video and continued fight for immigrant rights led to her visibility.

2. See Cristina Beltrán, "'Undocumented, Unafraid, and Unapologetic': DREAM Activists, Immigrant Politics, and the Queering of Democracy," in *From Voice to Influence: Understanding Citizenship in a Digital Age*, ed. Danielle Allen and Jennifer S. Light (Chicago: University of Chicago Press, 2015), 80–104.

3. Ibid., 81.

4. Ibid., 98.

5. Ibid., 81.

6. Ibid., 86.

7. Ibid., 94.

8. See Lauren Berlant, *The Queen of America Goes to Washington City: Essays on Sex and Citizenship* (Durham: Duke University Press, 1997), 62.

9. As quoted by Beltrán, "'Undocumented, Unafraid, and Unapologetic,'" 103.

10. See "Viridiana Martinez, North Carolina Dream Team," *Youtube.com*, April 5, 2011, https://www.youtube.com/watch?v=4Fbo4NT_p5M.

11. See Sarah Banet-Weiser, *Authentic™: The Politics of Ambivalence in a Brand Culture* (New York: New York University Press, 2012).

12. There is an odd parallel between Beltrán's description of undocumented activists and Banet-Weiser's description of neoliberal subjects engaged in practices of self-disclosure—of putting themselves out there transparently—that drive neoliberal conflations of branding with freedom. See Banet-Weiser's *Authentic*, 60.

13. See the script of Laura Poitras, *Citizenfour* (Praxis Films, 2014), http://www.springfieldspringfield.co.uk/movie_script.php?movie=citizenfour.

Chapter 4

1. Sara Ahmed, *The Cultural Politics of Emotion* (New York: Routledge, 2004), 45.

2. Jean-Luc Nancy, *The Inoperative Community*, trans. Peter Connor (Minneapolis: University of Minnesota Press, 1991), 67.

3. Eve Sedgwick, *Epistemology of the Closet* (Berkeley: University of California Press, 1990), 4.

4. ChiaVideos, "Amanda Todd's Story: Struggling, Bullying, Suicide, Self Harm," *YouTube.com*, October 11, 2012, https://www.youtube.com/watch?v=ej7afkypUsc.

5. Amanda Todd, "My Story: Struggling, Bullying, Suicide, Self Harm," *Youtube.com*, September 7, 2012, https://www.youtube.com/watch?v=vOHXGNx-E7E.

6. See Mark Kelley, "The Sextortion of Amanda Todd," *CBC: The Fifth Estate*, November 15, 2013, http://www.cbc.ca/fifth/episodes/2013-2014/the-sextortion-of -amanda-todd.

7. See Alye Pollack, "Words Do Hurt," *Youtube.com*, March 14, 2011, https:// www.youtube.com/watch?v=37_ncv79fLA.

8. See Jonah Mowry, "Jonah Mowry: 'Whats Goin On …,'" *Youtube.com*, August 10, 2011, https://www.youtube.com/watch?v=TdkNn3Ei-Lg.

9. See Kathleen Gasparini, "Suicide Prevention," *PBS.org* (n.d.), accessed October 24, 2015, http://www.pbs.org/inthemix/educators/lessons/depression2/; and Paul Bogush, "How to Make Notecard Confession Style Videos with Your Class …" *Edublogs.org*, February 11, 2013, http://blogush.edublogs.org/2013/02/11/how-to- make-note-card-confession-style-videos-with-your-class/.

10. See Sedgwick, "Epistemology of the Closet," in *Epistemology of the Closet*, 67–90.

11. As Sarah Banet-Weiser has shown, the line between authentic and illusory empowerment is muddy at best. Banet-Weiser, *Authentic™: The Politics of Ambivalence in a Brand Culture* (New York: New York University Press, 2012), 69.

12. "Was Bullied Teen's Video Fake?," *HLNTv.com*, February 8, 2012, http:// www.hlntv.com/video/2011/12/06/doctors-explain-why-teens-bully.

13. Quoted in ibid.

14. See Todd, "My Story."

15. Ibid.

16. See YiPing Tsou, "From 'Flash Mob' to 'Human Flesh Search,'" in *Digital AlterNatives with a Cause? Book 2—To Think*, ed. Nishant Shah and Fieke Jansen (Bangalore: Center for Internet Studies, 2011), 32–46.

17. ANONYMOUSOFFICIAL24, "Anonymous—Amanda Todd [Kody Maxson]," *YouTube.com*, October 19, 2012, http://youtu.be/_pv6PdhACfo.

18. See Danielle Allen, "Anonymous: On Silence and the Public Sphere," in *Speech and Silence in American Law*, ed. Austin Sarat (New York: Cambridge University Press, 2010), 106–133.

19. According to Silvan Tomkins, "shame is an experience of the self by the self": one can feel shame for others and on others' behalf, but shame is something experienced by a subject in the face of contempt. "Shame—Humiliation and Contempt— Disgust," in Tomkins, *Shame and Its Sisters: A Silvan Tomkins Reader*, ed. Eve Kosofsky Sedgwick and Adam Frank (Durham: Duke University Press, 1995), 136.

20. For more on this, see Lauren Berlant, "Chapter Two: Live Sex Acts (Parental Advisory: Explicit Material)," in Berlant, *The Queen of America Goes to Washington City: Essays on Sex and Citizenship* (Durham: Duke University Press, 1997), 55–82.

21. Tyler Kingkade, "Chelsea Chaney Sues Georgia School District for Using Facebook Photo of Her in a Bikini," *Huffington Post* (June 24, 2013), accessed August 31, 2014, http://www.huffingtonpost.com/2013/06/24/chelsea-chaney-facebook-bikini-photo_n_3487554.html.

22. Ibid.

23. See Wendy Chun and Sarah Friedland, "Habits of Leaking: Of Sluts and Network Cards," *Differences: A Journal of Feminist Cultural Studies* 26, no. 2 (2015): 1–28.

24. Tomkins, "Shame—Humiliation," 157.

25. Quoted in Camille Dodoro, "Hunter Moore on California's Revenge-Porn Law: 'You Fucking Retards,'" *Gawker.com*, October 4, 2013, http://gawker.com/hunter-moore-on-californias-revenge-porn-law-you-fuc-1440650903.

26. See Eden Osucha, "The Whiteness of Privacy: Race, Media, Law," *Camera Obscura* 24, no. 1 (2009): 67–107.

27. See Hortense Spillers, "Mama's Baby, Papa's Maybe: An American Grammar Book," *Diacritics* 17, no. 2 (1987): 65–81. In particular, she argues, "this profitable 'atomizing' of the captive body provides another angle on the divided flesh: we lose any hint or suggestion of a dimension of ethics, of relatedness between human personality and its anatomical features, between one human personality and another, between human personality and cultural institutions. To that extent, the procedures adopted for the captive flesh demarcate a total objectification, as the entire captive community becomes a living laboratory" (68).

28. Eva Cherniavsky, as quoted by Osucha, "The Whiteness of Privacy," 97.

29. Antonin Scalia, Syllabus of Supreme Court Decision in *Kyllo v. United States*, *Justia.com*, n.pag., June 11, 2001, https://supreme.justia.com/cases/federal/us/533/27/case.html.

30. See Tomkins, *Shame and Its Sisters*, 162. Tomkins argues: "the vicarious experience of shame, together with the vicarious experience of distress, is at once a measure of civilization and a condition of civilization. Shame enlarges the spectrum of objects outside of himself which can engage man and concern him. After having experienced shame through sudden empathy, the individual will never again be able to be entirely unconcerned with the other. But if empathy is a necessary condition for the development of personality and civilization alike, it is also a necessary condition for the experience of shame. If there is insufficient interest in the other, shame through empathy is improbable. How much shame can be felt at remediable conditions is one critical measure of the stage of development of any civilization."

31. Banet-Weiser, *Authentic*, 17.

32. See ibid., specifically 51–59, in "Branding the Postfeminist Self: The Labor of Femininity."

33. Ibid., 81.

34. Ibid., 89.

35. Tiqqun, *Preliminary Materials for a Theory of the Young-Girl*, trans. Ariana Reines (Los Angeles: Semiotext(e), 2012), 15.

36. Ibid., 65.

37. Ibid., emphasis in original, 101.

38. Tomkins, *Shame and Its Sisters*, 138.

39. Sedgwick, *Epistemology of the Closet*, 63.

40. Ibid., 11.

41. See Thomas Keenan, "Windows: Of Vulnerability," in *The Phantom Public Sphere*, ed. Bruce Robbins (Minneapolis: University of Minnesota Press, 1997), 121–141.

42. Dick Hebdige, quoted in E. Gabriella Coleman, "Phreaks, Hackers, and Trolls: The Politics of Transgression and Spectacle," in *The Social Media Reader*, ed. Michael Mandiberg (New York: New York University Press, 2012), 102.

43. Coleman, "Phreaks, Hackers, and Trolls," 102.

44. Michel Foucault, *The History of Sexuality*, vol. 1: *An Introduction*, trans. Robert Hurley (New York: Vintage Books, 1990), 45.

45. Jodi Dean, *Publicity's Secret: How Technoculture Capitalizes on Democracy* (Ithaca: Cornell University Press, 2002), 4.

46. Ibid., 45.

47. Ibid., 46. According to Dean, "to register as a subject in technoculture, one has to present oneself as an object for everyone else" (125).

48. Ibid., 124.

49. Ibid., 116, from "repeatedly trying, doing the same thing over and over and over again, even when, *especially when*, the actions are doomed to fail, is a pleasure in itself."

50. Ibid., 118. Desire (the conspiring subject) and drive (the celebrity subject) are caught in the twinning of publicity and surveillance.

51. See Keenan, "Windows: Of Vulnerability."

52. Dean, *Publicity's Secret*, 118.

53. Ibid., 127.

54. Ibid., 128.

55. Ibid., 124.

56. Keenan, "Windows: Of Vulnerability," 132.

57. See Dean, *Publicity's Secret*, 119–121.

58. Ibid., 124.

59. Tomkins, *Shame and Its Sisters*, 137.

60. Ibid., 134.

61. Ibid., 142.

62. D. A. Miller, *The Novel and the Police* (Berkeley: University of California Press, 1988), 213, 208.

63. Foucault, *History of Sexuality*, 1: 59–60.

64. Ibid., 60.

65. Ibid., emphasis in original, 35.

66. Tomkins, *Shame and Its Sisters*, 148, 133.

67. Ibid., 139.

68. Ibid., 57.

69. Ibid., 36.

70. Ibid., 139.

71. Ibid., 156–157.

72. Saidiya V. Hartman, *Scenes of Subjection: Terror, Slavery, and Self-Making in Nineteenth-Century America* (New York: Oxford University Press, 1997).

73. Ahmed, *Cultural Politics of Emotion*, 49.

74. Ibid., 51–52.

75. Ibid., 54.

76. Lisa Nakamura, "'It's a Nigger in Here! Kill The Nigger!': User-Generated Media Campaigns against Racism, Sexism, and Homophobia in Digital Games," in *The International Encyclopedia of Media Studies*, ed. Angharad Valdivia, vol. 6 (2013), http://lnakamur.files.wordpress.com/2013/04/nakamura-encyclopedia-of-media-studies-media-futures.pdf.

77. "Toronto 'Slut Walk' Takes to City Streets," *CBC News*, April 3, 2011, http://www.cbc.ca/news/canada/toronto/toronto-slut-walk-takes-to-city-streets-1.1087854.

78. Shilpa Phadke, Sameera Khan, and Shilpa Ranade, *Why Loiter? Women and Risk on Mumbai Streets* (New Delhi: Penguin Books, 2011), 60.

79. Ibid., 10.

80. As they note, "Public spaces—whether the Greek *agora* (marketplace), Roman fora, or American parks, commons, marketplaces and squares—were premised on exclusion even as they mediated interaction between people"—the modern city unbelongers are the poor. Ibid., 9.

81. Ibid., 19.

82. Ibid., 21.

83. Ibid., 188.

84. Ibid., 46.

85. Ibid., 178.

86. We would need to question why it is illegal in countries such as the United Kingdom to know what your Ethernet card knows. Packet sniffers do not automatically give access to encrypted information, but they do give users a sense of how their computers operate. Natalie Jeremijenko's early *Dangling String*, which reacted every time packets were sent across a local area network, demonstrates nicely the ways in which the rhythms of network exchange can be incorporated into our offline space. David C. Howe's and Helen Nissenbaum's *TrackMeNot*, a browser extension that randomly adds searches, is another way forward toward mass loitering.

87. Jean-Luc Nancy, *Corpus*, trans. Richard A. Rand (New York: Fordham University Press, 2008), 11.

88. Tomkins, *Shame and Its Sisters*, 137.

89. Ibid., 137.

90. See Jean-Luc Nancy, *Being Singular Plural*, trans. Robert D. Richardson and Anne E. O'Byrne (Stanford: Stanford University Press, 2000).

91. See *Project Unbreakable*, http://projectunbreakable.tumblr.com/, accessed October 26, 2015, and *We Are the 99 Percent*, http://wearethe99percent.tumblr.com/, last modified October 14, 2013.

92. See Nancy, *Being Singular Plural*, 5.

93. Ibid., 3.

94. Jean-Luc Nancy, "Exscription," trans. Katherine Lydon, *Yale French Studies* 78 (1990): 63.

95. Ibid.

96. Ibid., 64.

97. Ibid., 50.

98. Nancy, *The Inoperative Community*, 61.

Reality ≠ Truth

1. See Associated Press, "Gadget Addicts Feel 'Phantom Vibrations,'" *CBS News*, October 11, 2007, http://www.cbsnews.com/news/gadget-addicts-feel-phantom-vibrations/; and Associated Press, "Cell-Phone Junkies Feel Phantom Ring Vibrations," *Fox News*, October 12, 2007, http://www.foxnews.com/story/0,2933,301172,00.html.

2. See Michelle Drouin, Daren H. Kaiser, and Daniel A. Miller, "Phantom Vibrations among Undergraduates: Prevalence and Associated Psychological Characteristics," *Computers in Human Behavior* 28, no. 4 (2012): 1490–1496.

3. Associated Press, "Gadget Addicts."

4. Drouin, Kaiser, and Miller, "Phantom Vibrations among Undergraduates."

5. See V. S. Ramachandran and Eric L. Altschuler, "The Use of Visual Feedback, in Particular Mirror Visual Feedback, in Restoring Brain Function," *Brain* 132, no. 7 (2009): 1693–1710. No doubt, the notion of mirror neurons—and the possible cortical mapping of sensations—played a part in the conception of phantom vibration syndrome, since they made possible the view of the human as cortically attached to missing limbs.

6. Félix Ravaisson, *Of Habit*, trans. Clare Carlisle and Mark Sinclair (London: Continuum, 2008).

7. M. B. Rothberg, A. Arora, J. Hermann, R. Kleppel, P. St Marie, and P. Visintainer, "Phantom Vibration Syndrome among Medical Staff: A Cross Sectional Survey," *BMJ* 341 (2010): c6914.

Conclusion

1. Natalie Bookchin, *Mass Ornament* (2009): https://vimeo.com/5403546; *Testament* (2009): http://bookchin.net/projects/testament; both accessed October 28, 2015.

2. Natalie Bookchin and Blake Stimson, "Out in Public: Natalie Bookchin in Conversation with Blake Stimson," *Video Vortex Reader II: Moving Images Beyond YouTube*, ed. Geert Lovink and Rachel Somers Miles (Amsterdam: Institute of Network Cultures, 2011), 308.

3. Jaimie Baron, "Subverted Intentions and the Potential for 'Found' Collectivity in Natalie Bookchin's *Mass Ornament*," *Maska: Performing Arts Journal* 26, nos. 143–144 (Winter 2011): 303–314.

4. See Natalie Bookchin, *Now he's out in public and everyone can see* (2012), accessed October 28, 2015, http://bookchin.net/projects/now-hes-out-in-public-and-everyone-can-see/.

5. Bookchin and Stimson, "Out in Public," 310.

Bibliography

"About the Framingham Heart Study." *Framingham Heart Study: A Project of the National Heart, Lung, and Blood Institute and Boston University*. Accessed January 17, 2016. https://www.framinghamheartstudy.org/about-fhs/index.php.

"Access to Certain Business Records for Foreign Intelligence and International Terrorism Investigations." U.S. Code Title 50, Chapter 36, Subchapter IV, § 1861. *Legal Information Institute*, Cornell University. Accessed October 20, 2015. https://www.law.cornell.edu/uscode/text/50/1861.

Agamben, Giorgio. *State of Exception*. Trans. Kevin Attell. Chicago: University of Chicago Press, 2005.

Agre, Philip E. "Surveillance and Capture: Two Models of Privacy." *Information Society* 10 (2) (1994): 101–127.

Ahmed, Sara. *The Cultural Politics of Emotion*. New York: Routledge, 2004.

Allen, Danielle. "Anonymous: On Silence and the Public Sphere," in *Speech and Silence in American Law*, ed. Austin Sarat (New York: Cambridge University Press, 2010), 106–133.

Althusser, Louis. *Lenin and Philosophy, and Other Essays*. London: New Left Books, 1971.

"Amanda Todd Suicide Response—My Own Story." *YouTube.com*. October 14, 2012. https://www.youtube.com/watch?v=EC0MOiOCcWA.

Anderson, Benedict. *Imagined Communities: Reflections on the Origin and Spread of Nationalism*. London: Verso, 1991.

Anderson, Chris. "The End of Theory: Will the Data Deluge Make the Scientific Method Obsolete?" *Edge*. N. pag. Jun. 30, 2009. http://edge.org/3rd_culture/anderson08/anderson08_index.html.

Ang, Ien. *Living Room Wars: Rethinking Media Audiences for a Postmodern World*. New York: Routledge, 1996.

ANONYMOUSOFFICIAL24. "Anonymous – Amanda Todd [Kody Maxson]" *YouTube.com*. October 19, 2012. http://youtu.be/_pv6PdhACfo.

Appadurai, Arjun. *Modernity at Large: Cultural Dimensions of Globalization*. Minneapolis: University of Minnesota Press, 1996.

Arendt, Hannah. "The Crisis in Education." In *Between Past and Future: Eight Exercises in Political Thought*. New York: Penguin, 1968.

Arendt, Hannah. *The Human Condition*. Chicago: University of Chicago Press, 1998.

Arrington, Michael. "85% of College Students Use FaceBook." *TechCrunch*. September 7, 2005. http://www.techcrunch.com/2005/09/07/85-of-college-students-use-facebook/.

Associated Press. "Cell-Phone Junkies Feel Phantom Ring Vibrations." *FOX News*. October 12, 2007. http://www.foxnews.com/story/0,2933,301172,00.html.

Associated Press. "Gadget Addicts Feel 'Phantom Vibrations.'" *CBS News*. October 11, 2007. http://www.cbsnews.com/news/gadget-addicts-feel-phantom-vibrations/.

"AT&T Whistle-Blower's Evidence." *Wired.com*. May 17, 2006. http://archive.wired.com/science/discoveries/news/2006/05/70908.

Axel, Brian Keith. "The Diasporic Imaginary." *Public Culture* 14 (2) (2002): 411–428.

Banet-Weiser, Sarah. *Authentic™: The Politics of Ambivalence in a Brand Culture*. New York: New York University Press, 2012.

Barlow, John Perry. "A Declaration of the Independence of Cyberspace," *EFF.org*. February 8, 1996. https://projects.eff.org/~barlow/Declaration-Final.html.

Baron, Jaimie. "Subverted Intentions and the Potential for 'Found' Collectivity in Natalie Bookchin's *Mass Ornament*." *Maska: Performing Arts Journal* no. 26: 143–44 (2011): 303–314.

Baudrillard, Jean. *In the Shadow of the Silent Majorities, or, The End of the Social, and Other Essays*. New York: Semiotext(e), 1983.

Baudrillard, Jean. "The Masses: The Implosion of the Social in the Media." Trans. Marie Maclean. *New Literary History* 16 (3) (1985): 577–589.

Beck, Ulrich. *Risk Society: Towards a New Modernity*. Trans. Mark Ritter. London: Sage, 1992.

Beltrán, Cristina. "'Undocumented, Unafraid, and Unapologetic': DREAM Activists, Immigrant Politics, and the Queering of Democracy." In *From Voice to Influence: Understanding Citizenship in a Digital Age*, ed. Danielle Allen and Jennifer S. Light, 80–104. Chicago: University of Chicago Press, 2015.

Benjamin, Walter. "The Story Teller." In *Illuminations*. New York: Harcourt, Brace Jovanovich, 1969.

Benkler, Yochai. *The Wealth of Networks: How Social Production Transforms Markets and Freedom*. New Haven: Yale University Press, 2006.

Bennett, T., F. Dodsworth, G. Noble, M. Poovey, and M. Watkins. "Habit and Habituation: Governance and the Social." *Body and Society* 19 (2–3) (2013): 3–29.

Benveniste, Émile. *Problems in General Linguistics*. Trans. Mary Elizabeth Meek. Coral Gables, Fl: Miami University Press, 1971.

Berardi, Franco. *The Soul at Work: From Alienation to Autonomy*. Los Angeles: Semiotext(e), 2009.

Berlant, Lauren Gail. *Cruel Optimism*. Durham: Duke University Press, 2011.

Berlant, Lauren Gail. *The Queen of America Goes to Washington City: Essays on Sex and Citizenship*. Durham: Duke University Press, 1997.

Bernstein, Elizabeth. "Brokered Subjects and 'Illicit Networks.'" Invited lecture, Conference on Habits of Living, Brown University, Providence RI, 2013.

Bernstein, Elizabeth. *Temporarily Yours: Intimacy, Authenticity, and the Commerce of Sex*. Chicago: University of Chicago Press, 2007.

Blackman, L. "Habit and Affect: Revitalizing a Forgotten History." *Body and Society* 19 (2–3) (2013): 186–216.

Blanchot, Maurice. *The Unavowable Community*. Trans. Pierre Joris. Barrytown, NY: Station Hill Press, 1988.

Bogush, Paul. "How to Make Notecard Confession Style Videos with your Class …" *Edublogs.org*. February 11, 2013. http://blogush.edublogs.org/2013/02/11/how-to -make-note-card-confession-style-videos-with-your-class/.

Bookchin, Natalie. *Mass Ornament*. 2009. Accessed October 28, 2015. https:// vimeo.com/5403546.

Bookchin, Natalie. *Now he's out in public and everyone can see*. 2012. Accessed October 28, 2015. http://bookchin.net/projects/now-hes-out-in-public-and-everyone -can-see/.

Bookchin, Natalie. *Testament*. 2009. Accessed October 28, 2015. http://bookchin.net/ projects/testament.

Bookchin, Natalie, and Blake Stimson. "Out in Public: Natalie Bookchin in Conversation with Blake Stimson." In *Video Vortex Reader II: Moving Images Beyond YouTube*, ed. Geert Lovink and Rachel Somers Miles, 306–317. Amsterdam: Institute of Network Cultures, 2011.

Bosker, Bianca. "Eric Schmidt on Privacy (VIDEO): Google CEO Says Anonymity Online Is 'Dangerous.'" *Huffington Post*, August 10, 2010. http://www.huffingtonpost .com/2010/08/10/eric-schmidt-privacy-stan_n_677224.html.

Bosker, Bianca. "Facebook's Randi Zuckerberg: Anonymity Online 'Has to Go Away.'" *Huffington Post*, July 27, 2011. http://www.huffingtonpost.com/2011/07/27/ randi-zuckerberg-anonymity-online_n_910892.html.

Bourdieu, Pierre. *Outline of a Theory of Practice*. Trans. Richard Nice. Cambridge: Cambridge University Press, 1977.

boyd, danah. "Friendster and Publicly Articulated Social Networks." Conference on Human Factors and Computing Systems (CHI 2004). ACM, Vienna, April 24–29, 2004. http://www.danah.org/papers/CHI2004Friendster.pdf.

boyd, danah, et al. "Friendship." In *Hanging Out, Messing Around, Geeking Out: Kids Living and Learning with New Media*, ed. Mizuko Ito et al., 79–84. Cambridge, MA: MIT Press, 2008.

boyd, danah, and Jeffrey Heer. "Profiles as Conversation: Networked Identity Performance on Friendster." Proceedings of the Hawai'i International Conference on System Sciences (HICSS-39). Persistent Conversation Trace, Kauai, HI: IEEE Computer Society. January 4–7, 2006. http://vis.stanford.edu/files/2006-Friendster-HICSS.pdf.

Brooks, Frederick P. *The Mythical Man-month: Essays on Software Engineering*. Reading, MA: Addison-Wesley Professional, 1995.

Brown, Wendy. *Undoing the Demos: Neoliberalism's Stealth Revolution*. Cambridge, MA: MIT Press, 2015.

Bruns, Axel. "At Times of Crisis, Twitter Shines Brightest." *Continuity* (Brisbane, Australia) (Q3 2012): 16–17.

Bruns, Axel. "Crisis Communication, Social Media, and the Environment." In *The Media and Communications in Australia*, ed. Stuart Cunningham and Sue Turnbull, 351–355. Sydney, Australia: Allen and Unwin, 2014.

Burkitt, Ian. "Technologies of the Self: Habitus and Capacities." *Journal for the Theory of Social Behaviour* 32 (2) (2002): 219–237.

Butler, Judith. *Excitable Speech: A Politics of the Performative*. New York: Routledge, 1997.

Bynum, Terrell Ward. "Anonymity on the Internet and Ethical Accountability." *The Research Center on Computing and Society*. 1997. Posted September 5, 2010. http:// rccs.southernct.edu/on-the-emerging-global-information-ethics/.

Carbone, Nick. "Top 10 Viral Videos." *Time.com*, December 4, 2012. http:// entertainment.time.com/2012/12/04/top-10-arts-lists/slide/kony-2012/.

Carlisle, Clare. "Creatures of Habit: The Problem and the Practice of Liberation." *Continental Philosophy Review* 38 (1–2) (2005): 19–39.

Castells, Manuel. "The New Economy: Informationalism, Globalization, Networking." In *Rise of Network Society*, vol. 1 of *The Information Age: Economy, Society and Culture*, 77–162. New York: Blackwell, 2000.

Castoriadis, Cornelius. *The Imaginary Institution of Society*. Cambridge, MA: MIT Press, 1998.

de Certeau, Michel. *The Practice of Everyday Life*. Trans. S. Rendall. Berkeley: University of California Press, 1984.

Chiang, Mung. *Networked Life: 20 Questions and Answers*. Cambridge: Cambridge University Press, 2012.

ChiaVideos. "Amanda Todd's Story: Struggling, Bullying, Suicide, Self Harm." *YouTube.com*, October 11, 2012. https://www.youtube.com/watch?v=ej7afkyp Usc.

Christakis, N. A., and J. H. Fowler. "The Spread of Obesity in a Large Social Network over 32 Years." *New England Journal of Medicine* 357 (4) (2007): 370–379.

Christensen, Brett M. "Osama Bin Laden Virus Emails." *Hoax-Slayer*, July 2, 2005. http://www.hoax-slayer.com/bin-laden-captured.html.

Chun, Wendy Hui Kyong. *Control and Freedom: Power and Paranoia in the Age of Fiber Optics*. Cambridge, MA: MIT Press, 2006.

Chun, Wendy Hui Kyong. "The Enduring Ephemeral, or The Future Is a Memory." *Critical Inquiry* 35 (1) (2008): 148–171.

Chun, Wendy Hui Kyong. "Hypo-Real Models, or Global Climate Change: A Challenge for the Humanities." *Critical Inquiry* 41 (Spring 2015): 675–703.

Chun, Wendy Hui Kyong. "Nómades que imaginan." In *Nomadismos Tecnológicos*, ed. Jorge La Ferla and Giselle Beiguelman, 49–60. Buenos Aires: Espacio Fundación Telefónica/Instituto Sergio Motta, 2011.

Chun, Wendy Hui Kyong. *Programmed Visions: Software and Memory*. Cambridge, MA: MIT Press, 2011.

Chun, Wendy Hui Kyong, and Sarah Friedland. "Habits of Leaking: Of Sluts and Network Cards." *Differences: A Journal of Feminist Cultural Studies* 26, no. 2 (2015): 1–28.

Clough, Patricia Ticineto. *Autoaffection: Unconscious Thought in the Age of Teletechnology*. Minneapolis: University of Minnesota Press, 2000.

Coats, Daniel R. *Department of Justice Brief (Reno v. ACLU)*, January 21, 1997. http://www.ciec.org/SC_appeal/970121_DOJ_brief.html.

Cohen, Jennifer. "10 Morning Habits Successful People Swear By." *Forbes*, November 26, 2014. http://www.forbes.com/sites/jennifercohen/2014/11/26/10-morning -habits-successful-people-swear-by/.

Coleman, Gabriella. "Our Weirdness Is Free: The Logic of Anonymous—Online Army, Agent of Chaos, and Seeker of Justice." *Triple Canopy* 15 (January 13, 2012). http://www.canopycanopycanopy.com/contents/our_weirdness_is_free.

Coleman, Gabriella. "Phreaks, Hackers, and Trolls: The Politics of Transgression and Spectacle." In *The Social Media Reader*, ed. Michael Mandiberg, 99–119. New York: New York University Press, 2012.

Coleman, Gabriella. "The Politics of Rationality: Psychiatric Survivors' Challenge to Psychiatry." In *Tactical Biopolitics: Art, Activism, and Technoscience*, ed. Beatriz da Costa and Kavita Philip, 341–363. Cambridge, MA: MIT Press, 2008.

Cooley, Heidi Rae. *Finding Augusta: Habits of Mobility and Governance in the Digital Era*. Hanover: Dartmouth College Press, 2014.

Covey, Stephen R. *7 Habits of Highly Effective People*. New York: Free Press, 1989.

Dash, Anil. *Twitter*. March 17, 2014. https://twitter.com/anildash/status/ 445761831597273088.

Dean, Jodi. *Publicity's Secret: How Technoculture Capitalizes on Democracy*. Ithaca: Cornell University Press, 2002.

Deleuze, Gilles. *Difference and Repetition*. Trans. Paul Patton. New York: Columbia University Press, 1994.

Deleuze, Gilles. *Empiricism and Subjectivity: An Essay on Hume's Theory of Human Nature*. Trans. Constantin V. Boundas. New York: Columbia University Press, 1991.

Deleuze, Gilles, and Félix Guattari. *A Thousand Plateaus: Capitalism and Schizophrenia*. Trans. Brian Massumi. Minneapolis: University of Minnesota Press, 1987.

Derrida, Jacques. "Cogito and the History of Madness." In *Writing and Difference*, trans. Alan Bass. Chicago: University of Chicago Press, 1978.

Derrida, Jacques. "Force of Law: The Mystical Foundation of Authority." In *Acts of Religion*, ed. Gil Anidjar, 228–298. New York: Routledge, 2002.

Derrida, Jacques. "Plato's Pharmacy." In *Dissemination*, trans. Barbara Johnson, 61–171. Chicago: University of Chicago Press, 1981.

Derrida, Jacques. *Politics of Friendship*. London: Verso, 1997.

Dewey, John. *Human Nature and Conduct: An Introduction to Social Psychology*. New York: Henry Holt, 1922.

Dienst, Richard. *Still Life in Real Time: Theory after Television*. Durham: Duke University Press, 1994.

Doane, Mary Ann. "Information, Crisis, Catastrophe." In *New Media, Old Media: A History and Theory Reader*, ed. Wendy Hui Kyong Chun and Thomas Keenan, 251–264. New York: Routledge, 2006.

Dodoro, Camille. "Hunter Moore on California's Revenge-Porn Law: 'You Fucking Retards.'" *Gawker.com*, October 4, 2013. http://gawker.com/hunter-moore-on -californias-revenge-porn-law-you-fuc-1440650903.

Draznin, Haley. "New York to Pay $17.9 Million to 2004 Republican Convention Protestors." *CNN.com*, January 16, 2014. http://edition.cnn.com/2014/01/15/ politics/new-york-republican-convention-settlement/.

Drouin, Michelle, Daren H. Kaiser, and Daniel A. Miller. "Phantom Vibrations among Undergraduates: Prevalence and Associated Psychological Characteristics." *Computers in Human Behavior* 28 (4) (2012): 1490–1496.

Duhigg, Charles. "How Companies Learn Your Secrets." *New York Times*, February 16, 2012. http://www.nytimes.com/2012/02/19/magazine/shopping-habits.html? pagewanted=all.

Duhigg, Charles. *The Power of Habit: Why We Do What We Do in Life and Business*. New York: Random House, 2012.

Edwards, James R., Jr. "Illegal Alien Heads Obamacare Navigator Program in NYC." *Center for Immigration Studies*, October 15, 2013. http://cis.org/edwards/illegal-alien -heads-obamacare-navigator-program-nyc.

Ernst, Wolfgang. "Dis/continuities: Does the Archive Become Metaphorical in Multi-Media Space?" In *New Media, Old Media: A History and Theory Reader*, ed. Wendy Hui Kyong Chun and Thomas Keenan, 105–124. New York: Routledge, 2006.

Feuer, Jane. "The Concept of Live Television: Ontology as Ideology." In *Regarding Television: Critical Approaches*, ed. E. A. Kaplan, 12–22. Washington, DC: American University Press, 1983.

Foderaro, Lisa. "Private Moment Made Public, Then a Fatal Jump." *New York Times*, September 29, 2010. http://www.nytimes.com/2010/09/30/nyregion/30suicide. html?_r=3&.

Foucault, Michel. *The Birth of Biopolitics: Lectures at the Collège de France, 1978–1979*. Ed. M. Senellart. Trans. Graham Burchell. New York: Palgrave Macmillan, 2008.

Foucault, Michel. *Discipline and Punish: The Birth of the Prison*. Trans. A. Sheridan. New York: Vintage, 1995.

Foucault, Michel. *The History of Sexuality*. Vol. 1, *An Introduction*. Trans. Robert Hurley. New York: Vintage, 1990.

Foucault, Michel. "What Is Critique?" In *What Is Enlightenment? Eighteenth-Century Answers and Twentieth-Century Questions*, ed. James Schmidt, 382–398. Berkeley: University of California Press, 1996.

Freud, Sigmund. "A Note upon the 'Mystic Writing Pad.'" In *General Psychological Theory: Papers on Metapsychology*, ed. Philip Rieff, 211–216. New York: Collier, 1963.

Friedberg, Anne. *The Virtual Window: From Alberti to Microsoft*. Cambridge, MA: MIT Press, 2006.

Friedman, Milton. *Capitalism and Freedom*. Chicago: University of Chicago Press, 1982.

Frizell, Sam. "The Missing Malaysian Plane: 5 Conspiracy Theories." *Time.com*, March 11, 2014. http://time.com/20351/missing-plane-flight-mh370-5-conspiracy -theories/.

Frohne, Ursula. "Screen Tests: Media, Narcissism, Theatricality, and the Internalized Observer." In *CTRL [Space]: Rhetorics of Surveillance from Bentham to Big Brother*, ed. Thomas Levin, Ursula Frohne, and Peter Weibel, 252–277. Cambridge, MA: MIT Press, 2002.

"Fueling Long-Term Impact." *TeachforAmerica.com*. n.p. n.d. https://www .teachforamerica.org/about-us/our-mission/our-impact.

Galloway, Alexander R. *The Interface Effect*. Cambridge, UK: Polity, 2012.

Galloway, Alexander R. *Protocol: How Control Exists after Decentralization*. Cambridge, MA: MIT Press, 2004.

Gasparini, Kathleen. "Suicide Prevention." *PBS.org*. Accessed October 24, 2015. http://www.pbs.org/inthemix/educators/lessons/depression2/.

Gates, Bill, Nathan Myhrvold, and Peter Rinearson. *The Road Ahead*. New York: Viking, 1995.

Giacobbe, Alyssa. "Who Failed Phoebe Prince?" *Boston Magazine*, June 2012. http:// www.bostonmagazine.com/2010/05/phoebe-prince/.

Gibson, William. *Neuromancer*. New York: Ace Books, 1984.

Gitelman, Lisa. *"Raw Data" Is an Oxymoron*. Cambridge, MA: MIT Press, 2013.

Goddard, Alexandra. "I am the Blogger who Allegedly 'Complicated' the Steubenville Gang Rape Case—and I Wouldn't Change a Thing." *Xo Jane*, March 18, 2013. http://www.xojane.com/issues/steubenville-rape-verdict-alexandria-goddard.

"Godwin's Law." *Wikipedia.com*. N. pag. Accessed August 27, 2014. http://en .wikipedia.org/wiki/Godwin's_law.

Goldberg, Michelle. "New York Lockdown." *Guardian*, August 11, 2004. http:// www.theguardian.com/politics/2004/aug/11/education.

Gore, Al. "Information Superhighways Speech." *International Telecommunications Union*, March 21, 1994. http://vlib.iue.it/history/internet/algorespeech.html.

Goyal, Sanjeev. *Connections: An Introduction to the Network Economy*. Princeton: Princeton University Press, 2007.

Granovetter, Mark S. "The Strength of Weak Ties." *American Journal of Sociology* 78 (6) (1973): 1360–1380.

Graybiel, A. M. "The Basal Ganglia and Chunking of Action Repertoires." *Neurobiology of Learning and Memory* 70 (1–2) (1998): 119–136.

Greenwald, Glenn. "NSA Collecting Phone Records of Millions of Verizon Customers Daily." *Guardian*, June 6, 2013. N. pag. http://www.theguardian.com/world/2013/jun/06/nsa-phone-records-verizon-court-order.

Grosz, Elizabeth. "Habit Today: Ravaisson, Bergson, Deleuze and Us." *Body and Society* 19 (2–3) (2013): 217–239.

Habermas, Jürgen. *The Structural Transformation of the Public Sphere: An Inquiry into a Category of Bourgeois Society*. Trans. Thomas Burger and Frederick Lawrence. Cambridge, MA: MIT Press, 1991.

Haraway, Donna Jeanne. *Simians, Cyborgs, and Women: The Reinvention of Nature*. New York: Routledge, 1991.

Hartman, Saidiya V. *Scenes of Subjection: Terror, Slavery, and Self-Making in Nineteenth-Century America*. New York: Oxford University Press, 1997.

Harvey, David. *A Brief History of Neoliberalism*. Oxford: Oxford University Press, 2005.

Heidegger, Martin. *Being and Time: A Translation of Sein und Zeit*. Trans. J. Stambaugh. Albany: State University of New York Press, 1996.

Heidegger, Martin. *The Question Concerning Technology, and Other Essays*. New York: Harper and Row, 1977.

Hirsch, Tad, and John Henry. "TXTmob: Text Messaging for Protest Swarms." In *CHI '05 Extended Abstracts on Human Factors in Computing Systems (CHI EA '05)*, 1455–1458. New York: ACM, 2005.

Hodgson, Geoffrey M. "The Ubiquity of Habits and Rules." *Cambridge Journal of Economics* 21 (6) (1997): 663–684.

Hume, David. *A Treatise of Human Nature*. Ed. David Fate Norton and Mary J. Norton. Oxford: Oxford University Press, 2000.

Hyun-Kyong, Kang. "Cell Phones Create Youth Nationalism." *Korea Times Online*. May 12, 2008. http://koreatimes.co.kr/www/news/special/2008/06/180_24035.html.

Jacob Abrams, et al. v. United States. 250 U.S. 616. J. Holmes, dissenting. 1919.

Jaguar, Jet. "Bizarre Photos and Quotes from Protestors at RNC Convention (Vanity)." *FreeRepublic.com*, August 29, 2004. Last modified September 9, 2004 http://www.freerepublic.com/focus/news/1202228/posts.

James, William. "Habit." *The Principles of Psychology*. University of Michigan Library, 1890. http://psychclassics.asu.edu/James/Principles/prin4.htm.

Jameson, Fredric. "Cognitive Mapping." In *Marxism and the Interpretation of Culture*, ed. Cary Nelson and Lawrence Grossberg, 347–357. Urbana: University of Illinois Press, 1988.

Jameson, Fredric. *Postmodernism, or, The Cultural Logic of Late Capitalism*. Durham: Duke University Press, 1991.

Jin, Y., B. F. Liu, and L. L. Austin. "Examining the Role of Social Media in Effective Crisis Management: The Effects of Crisis Origin, Information Form, and Source on Public's Crisis Responses." *Communication Research* 41 (1) (2011): 74–94.

Johnson, Barbara. "Apostrophe, Animation, and Abortion." *Diacritics* 16 (1) (1986): 29–47.

Johnson, Barbara. Translator's Introduction. In Jacques Derrida, *Dissemination*. Chicago: University of Chicago Press, 1981.

Johnston, Casey. "Netflix Never Used Its $1 Million Algorithm Due to Engineering Costs." *Wired*, April 16, 2012. http://www.wired.com/2012/04/netflix-prize-costs/.

Kahney, Leander. "E-mail Mobs Materialize All Over." *Wired*, July 5, 2003. http://archive.wired.com/culture/lifestyle/news/2003/07/59518.

Kandel, Eric R. *In Search of Memory: The Emergence of a New Science of Mind*. New York: W.W. Norton, 2006.

Kant, Immanuel. "An Answer to the Question: What Is Enlightenment?" In *What Is Enlightenment? Eighteenth-Century Answers and Twentieth-Century Questions*, ed. James Schmidt, 58–64. Berkeley: University of California Press.

Keay, Douglas. "Margaret Thatcher Interview ('No Such Thing as Society')," *Woman's Own*, September 23, 1987. N. pag. http://www.margaretthatcher.org/speeches/displaydocument.asp?docid=106689.

Keenan, Thomas. *Fables of Responsibility: Aberrations and Predicaments in Ethics and Politics*. Stanford: Stanford University Press, 1997.

Keenan, Thomas. "Publicity and Indifference (Sarajevo on Television)." *PMLA* 117 (1) (2002): 104–116.

Keenan, Thomas. "Windows: Of Vulnerability." In *The Phantom Public Sphere*, ed. Bruce Robbins, 121–141. Minneapolis: University of Minnesota Press, 1997.

Kelley, Mark. "The Sextortion of Amanda Todd." *CBC: The Fifth Estate*, November 15, 2003. http://www.cbc.ca/fifth/episodes/2013-2014/the-sextortion-of-amanda-todd.

Kimpel, Amy. "Using Laws Designed to Protect as a Weapon: Prosecuting Minors under Child Pornography Laws." *New York University Review of Law and Social Change* 34 (2010): 299.

Kingkade, Tyler. "Chelsea Chaney Sues Georgia School District for Using Facebook Photo of Her in a Bikini." *Huffington Post*, June 24, 2013. http://www.huffingtonpost .com/2013/06/24/chelsea-chaney-facebook-bikini-photo_n_3487554.html.

Kirschenbaum, Matthew G. *Mechanisms: New Media and the Forensic Imagination.* Cambridge, MA: MIT Press, 2008.

Klein, Naomi. *The Shock Doctrine: The Rise of Disaster Capitalism.* New York: Picador, 2007.

Kolata, Gina. "Catching Obesity from Friends May Not Be So Easy." *New York Times.* August 8, 2011. http://www.nytimes.com/2011/08/09/health/09network.html.

Kushner, David. "Anonymous vs. Steubenville." *Rolling Stone*, November 27, 2013. http://www.rollingstone.com/culture/news/anonymous-vs-steubenville-20131127.

Lacan, Jacques. "The Mirror Stage as Formative of the Function of the *I* as Revealed in Psychoanalytic Experience." In *Écrits: A Selection*, trans. A. Sheridan. London: Tavistock, 1977.

Landow, George P. *Hypertext: The Convergence of Contemporary Critical Theory and Technology.* Baltimore: Johns Hopkins University Press, 1992.

Latour, Bruno. *Reassembling the Social: An Introduction to Actor-Network-Theory.* Oxford: Oxford University Press, 2005.

"Leaked Steubenville Big Red Rape Video." *YouTube.com*, January 2, 2013. https:// www.youtube.com/watch?v=W1oahqCzwcY&bpctr=1400682132.

Lee, Justin. "The Reasonable Expectation of Invasion." *Science and Technology Law Quarterly* 1, no. 2. Accessed October 28, 2015. http://www.americanbar.org/ newsletter/publications/scitech_e_merging_news_home/privacy.html#_edn15.

Lehrer, Jonah. "Trails and Errors: Why Science Is Failing Us." *Wired*, December 16, 2011. N. pag. http://www.wired.com/2011/12/ff_causation/.

Lessig, Lawrence. *Code and Other Laws of Cyberspace.* New York: Basic Books, 1999.

Levin, Thomas. "Rhetoric of the Temporal Index: Surveillant Narration and the Cinema of 'Real Time.'" In *CTRL [Space]: Rhetorics of Surveillance from Bentham to Big Brother,* ed. Thomas Levin, Ursula Frohne, and Peter Weibel, 578–593. Cambridge, MA: MIT Press, 2002.

Levinas, Emmanuel. *Otherwise Than Being, or, Beyond Essence.* Trans. A. Lingis. Boston: Martinus Nijhoff, 1981.

Lévy, Pierre. *Collective Intelligence: Mankind's Emerging World in Cyberspace.* New York: Plenum Trade, 1997.

Liben-Nowell, D., and J. Kleinberg. "Tracing Information Flow on a Global Scale Using Internet Chain-letter Data." *Proceedings of the National Academy of Sciences of the United States of America* 105 (12) (2008): 4633–4638.

Loesch, Chris, and Keith Payne. "The Situated Inference Model of Priming: How a Single Prime Can Alter Perception, Goals, and Behavior." Talk given at the Society for Personality and Social Psychology Annual Meeting, Las Vegas, NV, January 2010.

Lovink, Geert. "Enemy of Nostalgia: Victim of the Present, Critic of the Future: Peter Lunenfeld Interviewed by Geert Lovink." *PAJ: A Journal of Performance and Art* 70 (2002): 5.

Lovink, Geert. *My First Recession: Critical Internet Culture in Transition*. Amsterdam: Institute of Network Cultures, 2011.

Lynch, Kevin. *The Image of the City*. Cambridge, MA: MIT Press, 1960.

Lyotard, Jean-François. *The Postmodern Condition: A Report on Knowledge*. Trans. G. Bennington and B. Massumi. Minneapolis: University of Minnesota Press, 1984.

Mackenzie, Adrian. *Cutting Code: Software and Sociality*. New York: Peter Lang, 2006.

Macur, Juliet, and Nate Schweber. "Rape Case Unfolds on Web and Splits City." *New York Times*, December 16, 2012. http://www.nytimes.com/2012/12/17/sports/high-school-football-rape-case-unfolds-online-and-divides-steubenville-ohio.html?pagewanted=all&_r=0.

Maha. "Protesting 101." *The Mahablog*, April 12, 2006. http://www.mahablog.com/2006/04/12/protesting-101/.

Mahoney, Michael. "The History of Computing in the History of Technology." *IEEE Annals of the History of Computing* 10 (2) (1988): 113–125.

Malabou, Catherine. "Addiction and Grace: Preface to Félix Ravaisson's *Of Habit*." In Félix Ravaisson, *Of Habit*, trans. C. Carlisle and M. Sinclair, vii–xx. London: Continuum, 2008.

Mankelow, Trent. "Quotes from Webstock 2012." *Optimal Usability Blog*. February 23, 2012. Accessed August 30, 2014. http://www.optimalusability.com/2012/02/quotes-from-webstock-2012/.

Marroquin, Maria. "Maria Marroquin, DreamActivist Pennsylvania." *Youtube.com*, April 5, 2011. https://www.youtube.com/watch?v=aFXTNk3YOnY.

Martinez, Viridiana. "Viridiana Martinez, North Carolina Dream Team." *Youtube.com*, April 5, 2011. https://www.youtube.com/watch?v=4Fbo4NT_p5M.

Massumi, Brian. *Parables for the Virtual: Movement, Affect, Sensation*. Durham, NC: Duke University Press, 2002.

Mayer-Schönberger, Viktor, and Kenneth Cukier. *Big Data: A Revolution That Will Transform How We Live, Work, and Think.* Boston: Houghton Mifflin Harcourt, 2013.

McDermott, K. B. "Implicit Memory." In *Encyclopedia of Psychology*, ed. Alan E. Kazdin, 4: 231–234. Washington, DC: American Psychological Association; Oxford: Oxford University Press, 2000.

McFadden, Robert. "City Is Rebuffed on the Release of '04 Records." *New York Times*, August 7, 2007. N. pag. http://www.nytimes.com/2007/08/07/nyregion/07police.html?ref=nationalspecial3.

McFadden, Robert. "Vast Anti-Bush Rally Greets Republicans in New York." *New York Times*, August 30, 2004. N. pag. http://www.nytimes.com/2004/08/30/politics/campaign/30protest.html.

McPherson, Tara. "Reload: Liveness, Mobility and the Web." In *The Visual Culture Reader*, 2nd ed., ed. Nicholas Mirzoeff, 458–473. New York: Routledge, 2004.

Merleau-Ponty, Maurice. *Phenomenology of Perception.* Trans. C. Smith. New York: Routledge, 2002.

Mieszkowski, Katherine. "Faking out Friendster." *Salon*, Aug. 13, 2013. http://www.salon.com/2003/08/14/fakesters/.

Mill, John Stuart. *On Liberty.* New York: Dover, 2002.

Miller, D. A. *The Novel and the Police.* Berkeley: University of California Press, 1988.

Mitchell, Kimberly J., David Finkelhor, Lisa M. Jones, and Janis Wolak. "Use of Social Networking Sites in Online Sex Crimes against Minors: An Examination of National Incidence and Means of Utilization." *Journal of Adolescent Health* 47 (2) (2010): 183–190.

"Most Highlighted Books of All Time." *Amazon.com*, n.d. Accessed September 9, 2013. https://kindle.amazon.com/most_popular/books_by_popular_highlights_all_time.

Mowry, Jonah. "Jonah Mowry: 'Whats Goin On … .'" *Youtube.com*, August 10, 2011. https://www.youtube.com/watch?v=TdkNn3Ei-Lg.

Moynihan, Colin. "City Subpoenas Creator of Text Messaging Code." *New York Times*, March 30, 2008. N. pag. http://www.nytimes.com/2008/03/30/nyregion/30text.html?ex=1364616000&en=aba124f5b62dae3c&ei=5124&partner=permalink&exprod=permalink.

Munster, Anna. *An Aesthesia of Networks: Conjunctive Experience in Art and Technology.* Cambridge, MA: MIT Press, 2013.

Nakamura, Lisa. "'It's a Nigger in Here! Kill the Nigger!': User-Generated Media Campaigns against Racism, Sexism, and Homophobia in Digital Games." *The*

International Encyclopedia of Media Studies, vol. 6. Ed. Angharad Valdivia. 2013. http://lnakamur.files.wordpress.com/2013/04/nakamura-encyclopedia-of-media-studies-media-futures.pdf.

Nancy, Jean-Luc. *Being Singular Plural*. Trans. Robert D. Richardson and Anne E. O'Byrne. Stanford: Stanford University Press, 2000.

Nancy, Jean-Luc. *Corpus*. Trans. R. Rand. New York: Fordham University Press, 2008.

Nancy, Jean-Luc. *The Experience of Freedom*. Trans. Bridget McDonald. Stanford: Stanford University Press, 1993.

Nancy, Jean-Luc. "Exscription." Trans. Katherine Lydon. *Yale French Studies* 78 (1990): 47–65.

Nancy, Jean-Luc. *The Inoperative Community*. Trans. P. Connor. Minneapolis: University of Minnesota Press, 1991.

Negt, Oskar, and Alexander Kluge. *Public Sphere and Experience: Toward an Analysis of the Bourgeois and Proletarian Public Sphere*. Trans. P. Labanyo and A. Oksiloff. Minneapolis: University of Minnesota Press, 1993.

"The Netflix Prize Rules." *Netflixprize.com*, 2009. http://www.netflixprize.com/rules.

Newman, M. E. J. *Networks: An Introduction*. Oxford: Oxford University Press, 2010.

Newman, M. E. J., Albert-László Barabási, and Duncan J. Watts, eds. *The Structure and Dynamics of Networks*. Princeton: Princeton University Press, 2006.

Nissenbaum, H. "Securing Trust Online: Wisdom or Oxymoron?" *Boston University Law Review* 81 (2001): 635–664.

Nissenbaum, H. *TrackMeNot*. Accessed January 17, 2016. https://cs.nyu.edu/trackmenot/.

Nitrozac & Snaggy. "On the Internet, Nobody Knows You're a Dog." *Joy of Tech*, 2013. http://www.geekculture.com/joyoftech/joyarchives/1862.html.

Noel, H., and B. Nyhan. "The 'Unfriending' Problem: The Consequences of Homophily in Friendship Retention for Causal Estimates of Social Influence." *Social Networks* 33 (3) (2011): 211–218.

Osucha, Eden. "The Whiteness of Privacy." *Camera Obscura* 24 (1) (2009): 67–107.

Phadke, Shilpa, Sameera Khan, and Shilpa Ranade. *Why Loiter? Women and Risk on Mumbai Streets*. New Delhi: Penguin Books, 2011.

Poitras, Laura. *Citizenfour*. Praxis Films, 2014. Script: http://www.springfieldspringfield.co.uk/movie_script.php?movie=citizenfour.

Pollack, Alye. "Words Do Hurt." *Youtube.com*, March 14, 2011. https://www.youtube.com/watch?v=37_ncv79fLA.

Povinelli, Elizabeth A. *The Empire of Love: Toward a Theory of Intimacy, Genealogy, and Carnality*. Durham: Duke University Press, 2006.

Project Unbreakable. Accessed October 26, 2015. http://projectunbreakable .tumblr.com/.

Ramachandran, V. S., and L. Eric Altschuler. "The Use of Visual Feedback, in Particular Mirror Visual Feedback, in Restoring Brain Function." *Brain* 132 (7) (2009): 1693–1710.

Rancière, Jacques. *Disagreement: Politics and Philosophy*. Trans. Julie Rose. Minneapolis: University of Minnesota Press, 1999.

Rancière, Jacques. *Dissensus: On Politics and Aesthetics*. Trans. S. Corcoran. London: Continuum, 2010.

Ravaisson, Félix. *Of Habit*. Trans. C. Carlisle and M. Sinclair. London: Continuum, 2008.

Reno v. ACLU. "Syllabus of Supreme Court Decision." June 26, 1997, n. pag. http:// www.ciec.org/SC_appeal/syllabus.shtml.

Ringrose, Jessica, Rosalind Gill, Sonia Livingstone, and Laura Harvey. *A Qualitative Study of Children, Young People and "Sexting": A Report Prepared for the NSPCC*. London: National Society for the Prevention of Cruelty to Children, 2012.

Robert, Mark, Christina Ray, Bryan Ware, and Gary Nan Tie. "Big Data, Systemic Risk and the U.S. Intelligence Community." *Enterprise Risk Management Symposium*, Chicago, April 22–24, 2013. N. pag. https://www.yumpu.com/en/document/ view/15445772/2013-chicago-erm-2f-mark.

Roediger, Henry. "Reconsidering Implicit Memory." In *Rethinking Implicit Memory*, ed. Jeffrey Bowers and Chad Marsolek. Oxford: Oxford University Press, 2003.

Rooney, Ben. "Big Data's Big Problem: Little Talent." *Wall Street Journal*, April 29, 2012. http://www.wsj.com/articles/SB10001424052702304723304577365700368073674.

Rooney, Ellen. "A Semiprivate Room." *Differences: A Journal of Feminist Cultural Studies* 13 (1) (2002): 128–156.

Roskies, A. L. "Are Neuroimages Like Photographs of the Brain?" *Philosophy of Science* 74 (5) (2007): 860–872.

Rothberg, M. B., A. Arora, J. Hermann, R. Kleppel, P. St Marie, and P. Visintainer. "Phantom Vibration Syndrome among Medical Staff: A Cross Sectional Survey." *BMJ (Clinical Research Ed.)* 341 (2010).

Rothberg, Peter. "Protest Pit Is a Dark, Shadowy Place." *Nation*, July 27, 2004. http:// www.thenation.com/blog/protest-pit-dark-shadowy-place#.

Rouvroy, Antoinette. "Technology, Virtuality and Utopia: Governmentality in an Age of Autonomic Computing." In *The Philosophy of Law Meets the Philosophy of Technology: Autonomic Computing and Transformations of Human Agency*, ed. Mireille Hildebrandt and Antoinette Rouvroy, 136–157. Milton Park, UK: Routledge, 2011.

Russell, Jason. *Kony 2012*. Invisible Children, Inc, 2012.

Scalia, Antonin. Syllabus of Supreme Court Decision in *Kyllo vs. United States*. June 11, 2001. *Justia.com*, n.pag. https://supreme.justia.com/cases/federal/us/533/27/case.html.

Schreber, Daniel Paul. *Memoirs of My Nervous Illness*. Trans. Ida Macalpine and Richard A. Hunter. New York: New York Review of Books, 2000.

Sedgwick, Eve Kosofsky. *Epistemology of the Closet*. Berkeley: University of California Press, 1990.

Sekula, Allan. "The Body and the Archive: The Use and Classification of Portrait Photography by the Police and Social Scientists in the Late 19th and Early 20th Centuries." *October* 39 (1986): 3–64.

Shalizi, Cosma, and Andrew Thomas. "Homophily and Contagion Are Generically Confounded in Observational Social Network Studies." *Sociological Methods and Research* 40 (2) (2011): 211–239.

Shannon, Claude Elwood, and Warren Weaver. *The Mathematical Theory of Communication*. Urbana: University of Illinois Press, 1949.

Shapiro, Rebecca. "Poppy Harlow, CNN Reporter, 'Outraged' Over Steubenville Rape Coverage Criticism: Report." *Huffington Post*, March 20, 2013. http://www.huffingtonpost.com/2013/03/20/poppy-harlow-cnn-steubenville-rape-coverage-criticism_n_2914853.html.

Shaviro, Steven. *Connected, or, What It Means to Live in the Network Society*. Minneapolis: University of Minnesota Press, 2003.

Shmueli, Sandra. "'Flash mob' Craze Spreads." *CNN.com*, August 8, 2003.http://www.cnn.com/2003/TECH/internet/08/04/flash.mob/index.html?_s=PM:TECH.

Simondon, Gilbert. "The Genesis of the Individual." Trans. Mark Cohen and Sanford Kwinter. In *Zone 6: Incorporations*, ed. Jonathan Crary and Sanford Kwinter, 296–319. New York: Zone, 1992.

Singel, Ryan. "Whistle-Blower Outs NSA Spy Room." *Wired*, April 7, 2006. http://archive.wired.com/science/discoveries/news/2006/04/70619.

"Slutwalk NYC 2011 Takes Over Union Square to Protest Slut-Shaming, Victim Blaming," *Huffington Post*, October 3, 2011. http://www.huffingtonpost.com/2011/10/03/slutwalk-nyc-2011-takes-o_n_992181.html#s384424.

Smith v. Maryland. 422 U.S. 735. 1979. Accessed October 20, 2015. http://caselaw.findlaw.com/us-supreme-court/442/735.html.

"Snowden Leaks: Google 'Outraged' at Alleged NSA Hacking." *BBC News*, October 31, 2013. http://www.bbc.com/news/world-us-canada-24751821.

Spillers, Hortense J. "Mama's Baby, Papa's Maybe: An American Grammar Book." *Diacritics* 17 (2) (1987): 65–81.

Tarde, Gabriel. *The Laws of Imitation.* Trans. W. C. P. Elsie. New York: H. Holt, 1903.

Tasch, Barbara. "Former Malaysian Leader Accuses CIA of Cover-Up in Missing Jet." *Time.com*, May 19, 2014. http://time.com/104480/malaysia-airliens-flight--mahathir-mohamad/.

Terranova, Tiziana. *Network Culture: Politics for the Information Age.* London: Pluto Press, 2004.

Tiqqun (collective). *Preliminary Materials for a Theory of the Young-girl.* Trans. Ariana Reines. Los Angeles: Semiotext(e), 2012.

Todd, Amanda. "My Story: Struggling, Bullying, Suicide, Self Harm." *Youtube.com*, September 7, 2012. https://www.youtube.com/watch?v=vOHXGNx-E7E.

Tomkins, Silvan. *Shame and Its Sisters: A Silvan Tomkins Reader.* Ed. Eve Kosofsky Sedgwick and Adam Frank. Durham: Duke University Press, 1995.

"Toronto 'Slut Walk' Takes to City Streets." *CBC News*, April 3, 2011. http://www.cbc.ca/news/canada/toronto/toronto-slut-walk-takes-to-city-streets-1.1087854.

Tsou, YiPing. "From 'Flash Mob' to 'Human Flesh Search.'" In *Digital AlterNatives with a Cause? Book 2—To Think*, ed. Nishant Shah and Fieke Jansen, 32–46. Bangalore: Center for Internet Studies, 2011.

Tulving, Endel. *Elements of Episodic Memory.* Oxford: Clarendon Press, 1983.

Turkle, Sherry. *Life on the Screen: Identity in the Age of the Internet.* New York: Simon and Schuster, 1995.

Verplanken, Bas, and Wendy Wood. "Interventions to Break and Create Consumer Habits." *Journal of Public Policy and Marketing* 25 (1) (2006): 90–103.

Virno, Paolo. "General Intellect." In *Lessico Postfordista.* Milan: Feltrinelli, 2001.

Von Neumann, John. "The First Draft Report on the EDVAC." *Contract No. W–670–ORD–4926* between the USORD and the University of Pennsylvania, June 30, 1945. http://www.virtualtravelog.net/wp/wp-content/media/2003-08-TheFirstDraft.pdf.

Wald, Priscilla. *Contagious: Cultures, Carriers, and the Outbreak Narrative.* Durham: Duke University Press, 2008.

Wark, McKenzie. "The Weird Global Media Event and the Tactical Intellectual." In *New Media, Old Media: A History and Theory Reader*, ed. Wendy Hui Kyong Chun and Thomas Keenan, 265–276. New York: Routledge, 2006.

"Was Bullied Teen's Video Fake?" *HLNTv.com*, February 8, 2012. http://www.hlntv.com/video/2011/12/06/doctors-explain-why-teens-bully.

Watts, Duncan J. *Six Degrees: The Science of a Connected Age.* New York: Norton, 2003.

We Are the 99 Percent. October 14, 2013. http://wearethe99percent.tumblr.com/.

Weiss, Gail. *Refiguring the Ordinary.* Bloomington: Indiana University Press, 2008.

Weisstein, Eric W. "Adjacency Matrix," *Wolfram MathWorld.* Accessed May 29, 2015. http://mathworld.wolfram.com/AdjacencyMatrix.html.

Weizenbaum, Joseph. *Computer Power and Human Reason: From Judgment to Calculation.* San Francisco: W. H. Freeman, 1976.

Wolak, Janis, David Finkelhor, and Kimberly J. Mitchell "Trends in Arrests for Child Pornography Production: The Third National Juvenile Online Victimization Study (NJOV-3)." April 2012. *Crimes against Children Research Center* (CV270). http://www.unh.edu/ccrc/pdf/CV270_Child%20Porn%20Production%20Bulletin_4-13-12.pdf.

Wolak, J., D. Finkelhor, K. J. Mitchell, and M. L. Ybarra. "Online 'Predators' and Their Victims: Myths, Realities, and Implications for Prevention and Treatment." *American Psychologist* 63 (2) (February–March 2008): 111–128.

Wood, Wendy, Jennifer S. Labrecque, Pei-Ying Lin, and Dennis Rünger. "Habits in Dual-Process Models." In *Dual-Process Theories of the Social Mind*, ed. Jeffrey W. Sherman, Bertram Gawronski, and Yaacov Trope, 371–385. New York: Guilford Press, 2014.

Wood, Wendy, and David T. Neal. "A New Look at Habits and the Habit-Goal Interface." *Psychological Review* 114 (4) (2007): 843–863.

Wood, Wendy, Jeffrey M. Quinn, and Deboarh A. Kashy. "Habits in Everyday Life: Thought, Emotion, and Action." *Journal of Personality and Social Psychology* 83 (6) (2002): 1281–1297.

Žižek, Slavoj. *First as Tragedy, Then as Farce.* London: Verso, 2009.

Žižek, Slavoj. *The Sublime Object of Ideology.* London: Verso, 1989.

Zuckerman, Ethan. "Cute Cat Theory Talk at Etech." *... My Heart's in Accra*, March 8, 2008. http://www.ethanzuckerman.com/blog/2008/03/08/the-cute-cat-theory-talk-at-etech/.

Index